SpringerBriefs in Applied Sciences and Technology

Continuum Mechanics

Series Editors

Holm Altenbach, Institut für Mechanik, Lehrstuhl für Technische Mechanik, Otto von Guericke University Magdeburg, Magdeburg, Sachsen-Anhalt, Germany

Andreas Öchsner, Faculty of Mechanical Engineering, Esslingen University of Applied Sciences, Esslingen am Neckar, Germany

These SpringerBriefs publish concise summaries of cutting-edge research and practical applications on any subject of Continuum Mechanics and Generalized Continua, including the theory of elasticity, heat conduction, thermodynamics, electromagnetic continua, as well as applied mathematics.

SpringerBriefs in Continuum Mechanics are devoted to the publication of fundamentals and applications, presenting concise summaries of cutting-edge research and practical applications across a wide spectrum of fields. Featuring compact volumes of 50 to 125 pages, the series covers a range of content from professional to academic.

More information about this subseries at http://www.springer.com/series/10528

Jingkai Chen

Nonlocal Euler–Bernoulli Beam Theories

A Comparative Study

 Springer

Jingkai Chen
College of Mechanical and Electronic Engineering
China University of Petroleum
Qingdao, China

ISSN 2191-530X   ISSN 2191-5318 (electronic)
SpringerBriefs in Applied Sciences and Technology
ISSN 2625-1329   ISSN 2625-1337 (electronic)
SpringerBriefs in Continuum Mechanics
ISBN 978-3-030-69787-7   ISBN 978-3-030-69788-4 (eBook)
https://doi.org/10.1007/978-3-030-69788-4

This Springer imprint is published by the registered company Springer Nature Switzerland AG
The registered company address is: Gewerbestrasse 11, 6330 Cham, Switzerland

This book is dedicated to my wife Xinyu Xiang who gives me unconditional love and my son Boyan Chen who keeps giving me random happiness.

Preface

The nonlocal theories can be dated back to 1950s when the local mechanics cannot explain the experimental observation in microscale. These classical nonlocal theories built a bridge between local continuum mechanics and molecular dynamics. Peridynamics is a nonlocal theory which is proposed by Silling in 2000. With the progress of the nanotechnology in recent years, the nonlocal beam theories attract the attention of researchers. Beam in microscale exhibits some complicated behavior due to the existence of long-range forces. In this book, a comparative study is proposed to discuss the static responses of the Euler-Bernoulli beams governed by different nonlocal beam theories, including the Eringen's stress-gradient beam theory, the Mindlin's strain-gradient beam theory, the higher-order beam theory, and the peridynamic beam theory. This book casts light on the fundamental understanding of the complicated nonlocal beam deformations in nano-electro-mechanical systems.

Qingdao, China   Jingkai Chen

Acknowledgements

This research is funded by National Natural Science Foundation of China (NO: 51804333), Postdoctoral Innovation Project of Shandong Province (sdbh20180010). The data that support the findings of this study are available from the author upon reasonable request. The author would like to thank Prof. Andreas Oechsner for his recommendation to SpringerBriefs in continuum mechanics.

Contents

Chapter 1
Introduction

The Euler–Bernoulli beam theory provides a formulation for describing the loading and deformation relation of beams in macroscale. It is proposed based on two assumptions: first, the beam deformation is small such that the linear elasticity theory holds; second, the cross sections of the beam remain plane and perpendicular to the neutral axis after the deformation. The first assumption provides a simple stress-strain linear relationship, while the second assumption introduces a certain strain-deformation relationship.

The Euler–Bernoulli beam equation was initially derived based on the local continuum mechanics, where the stress at one point is determined by the deformation of the other points in contact. The nonlocal beam theory was proposed after the appearance of nonlocal theories. Kroner derived the internal energy of materials with long-range cohesive forces and built the prototype of the nonlocal theory [1]. Eringen and Edelen proposed the nonlocal constitutive relations, which state that the stress at one point is determined not only by the strain of that point but also by the strains of points afar [2–5]. Based on the fact that nonlocal behaviors are often observed in small structures with the scale of molecules or atoms, the nonlocal beam theories attract much attentions in recent years. With the development of nanotechnology, the nonlocal beam theories are very often required to describe the behaviors of nano-scale beams. Recently, the beam theories based on the Eringen's nonlocal theory were proposed to describe the deformations of the nano-scale beams [6–8]. Aydogdu developed a generalized nonlocal beam theory to analyze the static and dynamic responses of single-walled carbon nanotubes [9]. Thai and Togun derived the nonlocal beam bending equation based on the Hamilton principle and the Eringen's nonlocal stress-strain relationship [10, 11]. Paola et al. rebuilt the Euler–Bernoulli beam model by introducing the concepts of long-range volume forces and moments [12]. Nonlocal beam theories based on the Eringen's theory can also be found in references [13, 14]. By using two principle kernel functions, a higher-order beam equation is derived based on the nonlocal strain gradient theory [15]. The nonlocal beam equations can also be derived from the modified nonlocal theories [16–18].

© The Author(s), under exclusive license to Springer Nature Switzerland AG 2021
J. Chen, *Nonlocal Euler–Bernoulli Beam Theories*,
SpringerBriefs in Continuum Mechanics,
https://doi.org/10.1007/978-3-030-69788-4_1

Peridynamics is a nonlocal theory, which describes material responses in integral form [19]. The early version of peridynamics is called bond-based peridynamics and has shown great advantages when modeling problems involving cracks [20]. A more powerful version of peridynamics called state based peridynamics was proposed and the limitations of bond-based peridynamics were overcome [21]. The nonlocal beam theories based on peridynamics was proposed in recent years. Moyer and Miraglia applied the bond based peridynamic concepts to represent the deformation of the Timoshenko beam [22]. O'Grady and Foster reformulated the Euler–Bernoulli beam equation based on the state based peridynamics [23]. The O'Grady-Foster's model reduces to the Euler–Bernoulli beam theory when the horizon reduces to zero, and is equivalent to the Eringen's nonlocal elasticity when the deformation is smooth. By comparing the bending moment calculated from traditional Euler–Bernoulli beam equation to the proposed nonlocal bending moment, the nonlocal influence function is found to be a function of the beam cross-section geometry. This non-ordinary state-based beam model is extended to describe the peridynamic plates and shells with arbitrary Poisson ratio [24, 25]. Diyaroglu et al. reformulated the beam equations by applying the principal of virtual work based on peridynamics [26]. The critical curvature and angles of the beams are derived from the critical energy release rate of the material, and the nonlocal dispersion relations is found to be much closer to real-materials than the classical dispersion relations. By utilizing Euler-Lagrange equations, the nonlocal Euler–Bernoulli beam formulation is derived based on state-based peridynamics [27]. Despite the fact that Diyaroglu's model is derived from different approach, the equation of motion is in the similar form as the O'Grady-Foster's model. Different types of nonlocal boundary conditions are discussed and numerical examples shows the validity of the peridynamic beam equations by comparing the nonlocal results with the classical Euler–Bernoulli deformations.

In the following chapters, the nonlocal beam theories are introduced, including the Eringen's stress-gradient beam equation, the Mindlin's strain gradient beam equation, the higher-order beam equation and the peridynamics beam equation. In particular, the peridynamic Euler–Bernoulli beam equation is reformulated based on bond-based peridynamics. The integral-form peridynamic beam equation is further expanded to be the infinite order differential equation by Tayler series expansion. An analytical solution to these nonlocal beam equations are derived and the nonlocal phenomenon is observed and discussed. Benchmark beam examples are solved analytically and numerically, including the simply-supported beam, the clamped-clampled beam and the cantilever beam. This book is organized as follows: Chap. 2 introduces the nonlocal beam equations based on the classical nonlocal theories. Chapter 3 introduces the peridynamics and the beam equation derived from the peridynamics. Chapter 4 derives the analytical solution of these nonlocal beam bending equations, including the simply-supported beam, the clamped-clamped beam and the cantilever beam. The numerical solutions to the integral-form peridynamic beam equation are proposed in Chap. 5. Finally, the conclusions are summarized in Chap. 6.

References

1. Kroner E (1967) Elasticity theory of materials with long range cohesive forces. Int J Solids Struct 3:731–742
2. Eringen AC (1972) Linear theory of nonlocal elasticity and dispersion of plane waves. Int J Eng Sci 10:425–435
3. Eringen AC (1972) Nonlocal polar elastic continua. Int J Eng Sci 10:1–16
4. Eringen AC (1983) On differential equations of nonlocal elasticity and solutions of screw dislocation and surface waves. J Appl Phys 54(9):4703–4710
5. Eringen AC, Edelen DGB (1972) On nonlocal elasticity. Int J Eng Sci 10:233–248
6. Wang CM, Kitipornchai S, Lim CW (2008) Beam bending solutions based on nonlocal timoshenko beam theory. J Eng Mech 134(6):475–481
7. Wang CM, Zhang YY, He XQ (2007) Vibration of nonlocal timoshenko beams. Nanotechnology 18(10):105401
8. Challamel N, Wang CM (2008) The small length scale effect for a nonlocal cantilever beam: a paradox solved. Nanotechnology 19(34):345703
9. Aydogdu M (2009) A general nonlocal beam theory: its application to nanobeam bending, buckling and vibration. Physica E 41:1651–1655
10. Thai H (2012) A nonlocal beam theory for bending, buckling, and vibration of nanobeams. Int J Eng Sci 52:56–64
11. Togun N (2016) Nonlocal beam theory for nonlinear vibrations of a nanobeam resting on elastic foundation. Bound Value Probl 1:1–14
12. Paola M, Failla G, Zingales M (2014) Mechanically based nonlocal euler bernoulli beam model. J Nanomechanics Micromech 4(1):A4013002
13. Ebrahimi F, Nasirzadeh P (2015) A nonlocal timoshenko beam theory for vibration analysis of thick nanobeam using differential transform method. J Theor Appl Mech 53(4):1041–1052
14. Piovan M, Filipich C (2014) Modeling of nonlocal beam theories for vibratory and buckling problems of nano-tubes. Mec Computacional XXXIII, 1601–1614
15. Lim CW, Zhang G, Reddy JN (2015) A higher-order nonlocal elasticity and strain gradient theory and its applications in wave propagation. J Mech Phys Solids 78:298–313
16. Apuzzo A, Barretta R, Faghidian SA, Luciano R, Marotti de Sciarra F (2018) Free vibrations of elastic beams by modified nonlocal strain gradient theory. Int J Eng Sci 133:99–108
17. Barretta R, Marotti de Sciarra F (2018) Constitutive boundary conditions for nonlocal strain gradient elastic nano-beams. Int J Eng Sci 130:187–198
18. Romano G, Barretta R (2017) Stress-driven versus strain-driven nonlocal integral model for elastic nano-beams. Compos B Eng 114:184–188
19. Silling S (2000) Reformulation of elasticity theory for discontinuities and long-range forces. J Mech Phys Solids 48:175–209
20. Gerstle W, Sau N, Silling S (2005) Peridynamic modeling of plain and reinforced concrete structures. SMiRT 18-B01-2. *Beijing*
21. Silling S, Epton M, Weckner O (2007) Peridynamic states and constitutive modeling. J Elast 88:151–184
22. Moyer ET, Miraglia MJ (2014) Peridynamic solution for Timoshenko beams. Engineering 6:304–317
23. O'Grady J, Foster J (2014) Peridynamic beams: a non-ordinary, state-based model. Int J Solids Struct 51:3177–3183
24. O'Grady J, Foster J (2014) Peridynamic plates and flat shells: a non-ordinary, state-based model. Int J Solids Struct 51:4572–4579
25. O'Grady J, Foster J (2016) A meshfree method for bending and failure in non-ordinary peridynamic shells. Comput Mech 57:921–929
26. Diyaroglu C, Oterkus E, Oterkus S, Madenci E (2015) Peridynamics for bending of beams and plates with transverse shear deformation. Int J Solids Struct 69–70:152–168
27. Diyaroglu C, Oterkus E, Oterkus S (2019) An Euler–Bernoulli beam formulation in an ordinary state-based peridynamic framework. Math Mech Solids 24(2):361–376

Chapter 2
Nonlocal Beam Equations

The nonlocal theories are dated back to 1960s when experiments had recognized the dispersion of elastic waves in the atomic lattices. This was explained by the existence of the long-range forces [1–4]. Classical local elasticity fails to describe this phenomenon. By introducing the nonlocal parameter, the nonlocal theories are proposed to describe phenomenon including the dispersion of elastic waves in lattices, the homogenization of composite, the behaviors of nematic elastomers and carbon nanotubes [5–8]. This chapter derives Euler–Bernoulli beam equations based on these classical nonlocal theories.

2.1 Eringen's Stress-Gradient Beam Equation

According to the Eringen's nonlocal theory, the stress at one point depends not only on the strain at the current point but also on strains of all other points inside the body [5, 6]. That is

$$\sigma_{ij} = \int_\Omega \gamma\left(\left|x' - x\right|, \delta_{eg}\right) C_{ijkl} \epsilon_{kl}\left(x'\right) dv\left(x'\right),\tag{2.1}$$

where γ is the nonlocal Modulus. The shape of the nonlocal Modulus γ is determined by the nonlocal parameter δ_{eg}. By choosing the Green's function of the proper linear differential operator as the nonlocal Modulus and operating on both side of the nonlocal equilibrium equation, the integral-form nonlocal stress can be expressed in differential form under static assumption [7]. The one-dimensional nonlocal stress can be reduced to

$$\sigma - \delta_{eg}^2 \sigma^{(2)} = E\epsilon\tag{2.2}$$

© The Author(s), under exclusive license to Springer Nature Switzerland AG 2021
J. Chen, *Nonlocal Euler–Bernoulli Beam Theories*,
SpringerBriefs in Continuum Mechanics,
https://doi.org/10.1007/978-3-030-69788-4_2

For the Euler–Bernoulli beam subjected to small deformation, the one-dimensional nonlocal stress at particular cross-section can be written as

$$\sigma - \delta_{eg}{}^2 \sigma^{(2)} = -Eyw^{(2)} \tag{2.3}$$

where δ_{eg} is the nonlocal parameter, E is the Young's Modulus, w is the transverse displacement of the neutral axis, y is the in-plane coordinate perpendicular to the reference beam axis. Then, the bending moment can be calculated by integrating nonlocal stress over certain cross-section. That is,

$$M = -\int_A \sigma y \, dA \tag{2.4}$$

By operating $\int_A [] \cdot y \, dA$ on both side of the Eq. (2.3), the bending equation is derived as

$$M - \delta_{eg}{}^2 M^{(2)} = EIw^{(2)} \tag{2.5}$$

For detailed process on above derivations, readers are referred to Ref. [9]. In this paper, we denote bending Eq. (2.5) as stress-gradient model since the one-dimensional stress-strain relation in Eq. (2.2) involves differential terms on stress.

2.2   Mindlin's Strain-Gradient Beam Equation

Similar to the Eringen's nonlocal theory in differential form, the nonlocal stress can be represented by higher-order differentiation on strain. Mindlin formulated the potential energy density of elastic solid as a function of strain and its first and second gradient. After rigorous derivation and simplification, the one-dimensional stress strain relation yields [1–3]

$$\sigma = E\left(\epsilon - \delta_{sg}{}^2 \epsilon^{(2)}\right) \tag{2.6}$$

Substitute Eqs. (2.6) to (2.4) yields an alternative beam bending equation. That is

$$M = EI\left(w^{(2)} - \delta_{eg}^2 w^{(4)}\right) \tag{2.7}$$

For detailed process on above derivations, readers are referred to Ref. [3]. In this paper, we denote bending Eq. (2.7) as strain-gradient model considering the fact that the one-dimensional stress-strain relation in Eq. (2.6) involves differential terms on strain.

2.3 Higher-Order Beam Equation

Instead of convoluting local strain with one kernel function, Lim et al. formulated the nonlocal stress-strain relation by introducing two principle kernel functions [10]. By combining the Eringen's nonlocal elasticity with strain gradient theory, a higher-order nonlocal elasticity which considers higher-order stress-gradient and strain-gradient nonlocality is proposed. Then, the simplified one-dimensional nonlocal stress-strain relation yields

$$\sigma - \delta_{eg}^2 \sigma^{(2)} = E\left(\epsilon - \delta_{sg}^2 \epsilon^{(2)}\right) \tag{2.8}$$

Substitute Eq. (2.8) to bending moment Eq. (2.4) yields an alternative beam bending equation. That is

$$M - \delta_{eg}^2 M^{(2)} = EI\left(w^{(2)} - \delta_{sg}^2 w^{(4)}\right) \tag{2.9}$$

For detailed process on above derivations, readers are referred to Ref. [10]. In this paper, we denote bending Eq. (2.9) as Higher-order model.

References

1. Mindlin RD, Tiersten HF (1961) Effects of couple-stresses in linear elasticity. Arch Ration Mech Anal 11(1):415–448
2. Mindlin RD (1963) Micro-structure in linear elasticity. Arch Ration Mech Anal 16(1):51–78
3. Mindlin RD (1965) Second gradient of strain and surface-tension in linear elasticity. Int J Solids Struct 1:417–438
4. Kroner E (1967) Elasticity theory of materials with long range cohesive forces. Int J Solids Struct 3:731–742
5. Eringen AC (1972) Linear theory of nonlocal elasticity and dispersion of plane waves. Int J Eng Sci 10:425–435
6. Eringen AC (1972) Nonlocal polar elastic continua. Int J Eng Sci 10:1–16
7. Eringen AC (1983) On differential equations of nonlocal elasticity and solutions of screw dislocation and surface waves. J Appl Phys 54(9):4703–4710
8. Eringen AC, Edelen DGB (1972) On Nonlocal Elasticity. Int J Eng Sci 10:233–248
9. Peddieson J, Buchanan GR, McNitt RP (2003) Application of nonlocal continuum models to nanotechnology. Int J Eng Sci 41:305–312
10. Lim CW, Zhang G, Reddy JN (2015) A higher-order nonlocal elasticity and strain gradient theory and its applications in wave propagation. J Mech Phys Solids 78:298–313

Chapter 3
Peridynamics Beam Equation

In nonlocal theories discussed in Chap. 2, the stress at one point is defined by integrating strain over a surrounding region. A nonlocal parameter is introduced to quantify the size of the region, and the nonlocal theories converge to the local elasticity as the nonlocal parameter goes to zero. Similar to local elasticity, these classical nonlocal theories are derived from the continuum assumption and ended up with the governing equations involving spatial differential terms. These spatial differential terms result in singularity when discontinuity appears. That is, both the local elasticity and nonlocal theories discussed previously are incapable of directly describing problems involve crack. Peridynamics is a new nonlocal theory which is formulated without continuum assumption [1]. Peridynamic equilibrium equation does not involves differential term, and it can be regarded as a continuous version of molecular dynamics or as a generalized version of local elastic theory. In this chapter, the earlier version of peridynamics, called bond-based peridynamics, and the advanced version of peridynamics, called state-based peridynamics, are introduced. Then the beam equations derived from peridynamics are discussed.

3.1 Bond-Based Peridynamics

The bond-based peridynamics was introduced by Silling [1]. Different from classical elasticity theory where the constitutive force acting on one point is only exerted by its direct contact points, the peridynamic constitutive force acting on one point is exerted by all points within the distance δ known as 'horizon'. The point located within its horizon region is called neighborhood. Peridynamics can be regarded as a continuous version of the molecular dynamics (MD). In fact, early peridynamics model utilized MD software LAMMPS to do numerical analysis. However, the constitutive law from peridynamics is calibrated differently from the Van der Waals force in MD. Peridynamics can also be regarded as a generalization of the local elasticity. But different

© The Author(s), under exclusive license to Springer Nature Switzerland AG 2021
J. Chen, *Nonlocal Euler–Bernoulli Beam Theories*,
SpringerBriefs in Continuum Mechanics,
https://doi.org/10.1007/978-3-030-69788-4_3

from the local elasticity which is built on continuum assumption, the peridynamics describes continuities and discontinuities in a consistent manner. Numerical implementations of the bond-based peridynamics were proposed [2–4]. Application of the bond-based peridynamic is limited to specific Poisson ratio, and shear deformation can not be captured.

The configuration of bond-based peridynamics is shown in Fig. 3.1. The point x located at the internal region (red dot in Fig. 3.1a) interacts to its neighboring points (blue dots in Fig. 3.1a) located within distance δ to point x. The horizon region of x is often expressed as $\mathcal{H}_x$. Each pair of particles within the same horizon interacts through a vector-valued function f called bond force. The bond force is often defined as a function of deformation and location of the interacting pair. As shown in Fig. 3.1b, the x and $\hat{x}$ are two points within the same horizon region. The bond force is expressed as

$$f = f\big(u(\hat{x}), u(x), \hat{x}, x\big) \tag{3.1}$$

The internal force acting on point x is calculated by integrating the bond force over its horizon. That is

$$\mathcal{L}_u(x) = \int_{\mathcal{H}_x} f\big(u(\hat{x}), u(x), \hat{x}, x\big) d\hat{V} \tag{3.2}$$

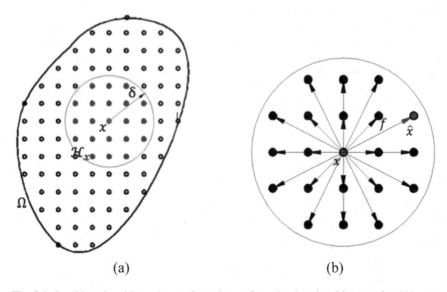

(a)   (b)

Fig. 3.1 Bond-based peridynamics configuration. **a** Over the domain of interest; **b** within one horizon region

Fig. 3.2 Deformation of
particle pair

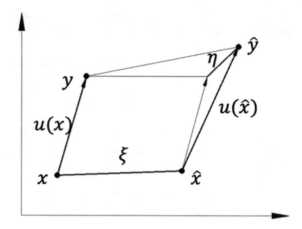

In Eqs. (3.1) and (3.2), the displacement of x and $\hat{x}$ are expressed as $u(x)$ and $u(\hat{x})$. The volume of $\hat{x}$ is expressed as $\hat{V}$. Note that the unit of peridynamic bond force is force per volume square, the unit of the internal force is force per volume. Then, the peridynamics equilibrium equation is formulated as

$$\rho \ddot{u}(x) = \mathcal{L}_u(x) + b(x) \tag{3.3}$$

The deformation configuration of the bond pair is shown in Fig. 3.2. The bond pair of particles x and $\hat{x}$ go to y and $\hat{y}$ respectively after deformation. u and $\hat{u}$ are displacements of the pair of particles. For simplification purpose, undeformed bond vector ξ and bond deformation vector η are introduced as show in Fig. 3.2. Those are

$$\xi = \hat{x} - x \tag{3.4}$$

$$\eta = u(\hat{x}) - u(x) \tag{3.5}$$

According to the bond deformation configuration in Fig. 3.2, the bond vector after deformation is

$$\xi + \eta = \hat{y} - y \tag{3.6}$$

Then the bond force can be expressed in a more compact form for homogeneous material. That is

$$f = f(\eta, \xi) \tag{3.7}$$

Note the Eq. (3.7) builds the relation between the deformation and the force in bond level, and is the peridynamic constitutive law. Therefore, the peridynamics describes the deformation of the body as deformation of the bond. The linear admissibility and angular admissibility conditions dictate the following relations:

$$f(\eta, \xi) = -f(-\eta, -\xi) \tag{3.8}$$

$$(\xi + \eta) \times f(\eta, \xi) = 0 \tag{3.9}$$

The linear admissibility condition ensures the bond forces between bond pair satisfy the Newton's third law while the angular admissibility condition is derived based on the assumption that the moment density is not exist at any point within the domain [1].

The linear admissibility condition and the angular admissibility condition further simplify the bond-based peridynamics constitutive relation as

$$f(\eta, \xi) = F(\eta, \xi)(\eta + \xi), \tag{3.10}$$

where $F(\eta, \xi)$ is a scalar valued function. The following relation is satisfied

$$F(\eta, \xi) = F(-\eta, -\xi), \tag{3.11}$$

The bond force between bond pair can be regarded as a spring connecting the center point and its neighborhood, as shown in Fig. 3.3.

Fig. 3.3 Bond force representation

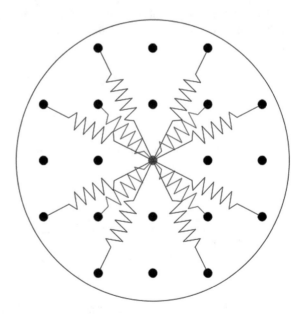

Note that the stiffness of the bond spring is nonlinear in general and is a scaler function of bond vector magnitude $|\boldsymbol{\xi}|$ for isotropic material. The constitutive relation in Eq. (3.10) can be linearized using Taylor series expansion. The linearized constitutive relation can be formulated by introducing the bond stiffness parameter c. That is

$$\boldsymbol{f}(\boldsymbol{\eta}, \boldsymbol{\xi}) = c(|\boldsymbol{\xi}|)s\boldsymbol{M}, \tag{3.12}$$

where

$$s = \frac{|\boldsymbol{\eta} + \boldsymbol{\xi}| - |\boldsymbol{\xi}|}{|\boldsymbol{\xi}|} \tag{3.13a}$$

$$\boldsymbol{M} = \frac{\boldsymbol{\eta} + \boldsymbol{\xi}}{|\boldsymbol{\eta} + \boldsymbol{\xi}|} \tag{3.13b}$$

In Eqs. (3.13a), (3.13b), s is the stretch of the bond which can be positive or negative depending on elongation or compression. $\boldsymbol{M}$ is the unit directional vector of the bond after deformation. Note that above linearization assumes no bond force at undeformed configuration.

The bond stiffness parameter is function of bond vector magnitude as shown in Fig. 3.4, it can be constant, triangular, inverse-triangular or any other functions as long as the peridynamic constitutive relation is consistent to the local elasticity. The consistent condition is often defined as the deformation energy consistency condition. That is, the deformation energy derived from peridynamics has to be equal to the deformation energy derived from local elasticity for uniform deformation.

Therefore, the linearized bond-based peridynamics equilibrium equation of isotropic material can be written as

$$\rho\ddot{\boldsymbol{u}}(x) = \int_{\mathcal{H}_x} c(|\boldsymbol{\xi}|)s\boldsymbol{M}d\hat{V} + \boldsymbol{b}(x) \tag{3.14}$$

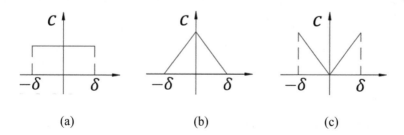

Fig. 3.4 Bond stiffness distribution. **a** Constant; **b** Triangular; **c** Inverse-Triangular

Fig. 3.5 Illustration of
boundary ambiguity

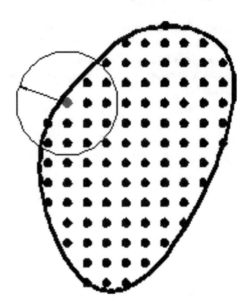

For particular discretization configuration, the discretized equilibrium equation
can be formulated using meshfree method [2]. Peridynamics as a nonlocal theory
has ambiguity problem on the boundary. This problem comes from the insufficient
representation of the neighboring region of boundary point, as shown in Fig. 3.5.
As a result, the 'stiffness' near the boundary is reduced. The boundary-induced
inaccuracy in peridynamics is called 'surface effect' or 'skin effect' and will be
discussed in Chap. 5.

3.2 State-Based Peridynamics

The bond-based peridynamics has mainly two limitations. First, the bond-based peridynamics can only describe deformation of material with certain Poisson ratio. That
is, 1/3 for two-dimensional plane stress problem and 1/4 for two-dimensional plane
strain and three-dimensional problems [5]. Second, the bond-based peridynamics
can not capture the shear deformation of the material. Therefore, the application
on the plastic deformation is limited. Then, the more advanced peridynamics, called
state-based peridynamics, is proposed [6]. The abovementioned two limitations were
overcome.

The state-based peridynamics is built on the mathematical concept of 'state'. A
state of order n is defined as an operator, which acts on vectors; the image of the
vector under this operator is an nth order tensor. The state is a nonlinear operator, the
detailed properties of the state operator is discussed [6]. Vector state is when $n = 1$
while scalar state is when $n = 0$. The operator of the vector state can be regarded as

Fig. 3.6 Configuration of
bond deformation state

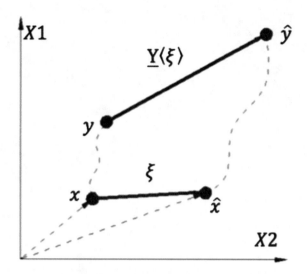

a special case of the second order tensor which is mathematically a linear operator.
Then, the concepts of bond deformation and bond force are represented using the
state operations, which are called 'force state' and 'deformation state'.

Deformation State. Like the strain tensor defined in local elasticity, the deformation
state is defined in state-based peridynamics as a vector state $\underline{Y}\langle.\rangle$. The configuration
of the deformation state is shown in Fig. 3.6. That is, the x and $\hat{x}$ are two point
within the same horizon before deformation. The pair of points x and $\hat{x}$ go to y and
$\hat{y}$ after deformation respectively. The deformation state of point x operates on the
undeformed bond vector ξ yields a new vector $\underline{Y}\langle\xi\rangle$. The image of the vector state
$\underline{Y}\langle\xi\rangle$ shows the bond vector after deformation.

The similarities and differences of the deformation state and the Green strain
tensor are summarized as follows:

- *The strain tensor operating on any unit vector yields to strain vector on the*
  corresponding normal surface; deformation state operates on any bond vectorξ
  gets extension of the corresponding bond.
- *The strain tensor at one point contains deformation information at all directions*
  of the point locally; while the deformation vector state at one point contains all
  deformation information between the point and all other point within its horizon.
- *The strain tensor can be regarded as a special form of the deformation state.*
  Mathematically, any strain tensor can be replaced by its corresponding defor-
  mation state while the reverse is not true. For example, deformation state can be
  nonlinear or even discontinuous, but strain tensor is a linear operator.

Force State. As the stress tensor defined in local elasticity, the force state defined in
state-based peridynamics is a vector state $\underline{T}\langle.\rangle$. The configuration of the force state
is shown in Fig. 3.7. That is, the x and $\hat{x}$ are two neighboring points. $\underline{T}[x]\hat{x} - x$ is

Fig. 3.7 Configuration of
bond force state

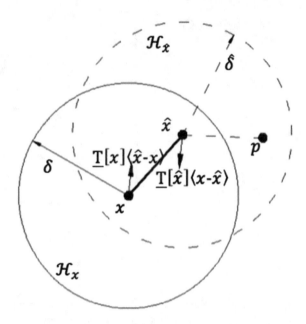

the image of force state $\underline{T}\langle.\rangle$ at point x operates on the bond vector $\hat{x} - x$. Similarly, $\underline{T}[\hat{x}]x - \hat{x}$ is the image of force state $\underline{T}\langle.\rangle$ at point $\hat{x}$ operates on the bond vector $x - \hat{x}$.

The state-based peridynamics defines the bond force between two pair points of the same neighborhood in the following form:

$$f(\hat{x}, x) = \underline{T}[x]\hat{x} - x - \underline{T}[\hat{x}]x - \hat{x}, \qquad (3.15)$$

Note that, in the general state-based peridynamics, the image of the force state does not have to be in the same direction as the bond vector, and the images of the force state of the pair points do not have to be in the same magnitude neither. The state-based peridynamics is called *ordinary* if the image vector of the force state is parallel to the bond vector during the deformation. Otherwise, the state-based peridynamics is called *nonordinary*. In a very special case that the image vector of the force state of the neighboring two points are both parallel to the bond vector and in the same magnitude, then the state-based peridynamics is reduced to the bond-based peridynamics. The differences of the bond-based peridynamics, the ordinary state-based peridynamics and the non-ordinary state-based peridynamics in bond force level are shown in Fig. 3.8.

Different from the bond-based peridynamics constitutive relation which correlates the bond force to the deformation of the corresponding bond, the state-based peridynamics constitutive relation correlates the force state to the deformation state. That is, the deformation state of certain point is determined by the deformation of all bonds within its neighborhood. Then, the bond force between pair points shown

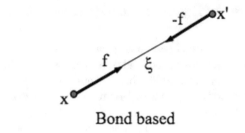

Bond based

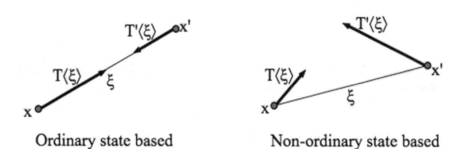

Ordinary state based   **Non-ordinary state based**

Fig. 3.8 Bond-force configurations [6]

in Eq. (3.15) is determined by the deformation state of both x and $\hat{x}$. As shown in Fig. 3.7, the bond force on x exerted by $\hat{x}$ is determined by the deformation of all neighboring points of x and the deformation of all neighboring points of $\hat{x}$. For example, point p is located within the horizon of $\hat{x}$ but is beyond the horizon of x, then the deformation of p has no influence on the bond force on x within the bond-based peridynamics framework but its deformation matters for state-based peridynamics.

Then, the equilibrium equation of the state-based peridynamics is expressed as

$$\rho\ddot{u}(x) = \int_{\mathcal{H}_x} \underline{T}[x]\hat{x} - x - \underline{T}[\hat{x}]x - \hat{x}d\hat{V} + b(x) \qquad (3.16)$$

Note that, the state-based peridynamic remedies the limitation of the bond-based peirdynamics. That is, the state-based peridynamics can simulate deformation of material with arbitrary Poisson ratio and shear deformation. Moreover, by introducing the concept of 'state', the force state and the deformation state in state-based peridynamics is much closer to the stress tensor and strain tensor in the local elasticity. Also, researches have shown that both bond-based peridynamics and state-based peridynamics are reduced to the local elasticity as the horizon size goes to zeros. The largest advantage of peridynamics over local elasticity or even the classical nonlocal theories is on the representation of discontinuous problems. However, the processing the peridynamics program is usually computational expensive.

3.3   Peridynamic Beam Equation Derivation

Inspired by the previously published peridynamic beam equations [7–12], the peri-dynamic beam equation in this paper is derived from bond-based peridynamics. For the Euler–Bernoulli beam deformation, the beam cross-section remains plane and perpendicular to the neutral axis during the deformation. Thus, only one independent variable w, which is the transverse displacement of the neutral axis, is sufficient to describe the deformation of the beam. Instead of defining deformation energy in bond level, the beam deformation energy density at certain cross-section is expressed as a function of nonlocal curvature. That is

$$W_{PD} = \frac{1}{2}\alpha\kappa_{PD}^2, \tag{3.17}$$

where κ_{PD} is the nonlocal curvature at point x and is defined based on small deformation assumption as

$$\kappa_{PD} = \int_{x-\delta}^{x+\delta} 2\beta(|\hat{x} - x|)\frac{\hat{w} - w}{(\hat{x} - x)^2}dx'. \tag{3.18}$$

To calibrate the values of parameters α and β, pure bending deformation is assumed with constant local curvature κ and a Cartesian coordinate is built such that the deformation at point x is zero, as shown in Fig. 3.9.

Therefore, the deformation of neighboring point $\hat{x}$ is written as

$$\hat{w} = \frac{1}{\kappa} - \sqrt{\left(\frac{1}{\kappa}\right)^2 - (\hat{x} - x)^2} \approx \frac{\kappa}{2}(\hat{x} - x)^2 \tag{3.19}$$

Considering the small deformation assumption, the deformation at neighboring point can be reduced to second order polynomial by Taylor series expansion up to second order. Then, substitutes to the peridynamic beam deformation density function yields

Fig. 3.9 Illustration of neutral axis after pure bending deformation in Cartesian coordinate

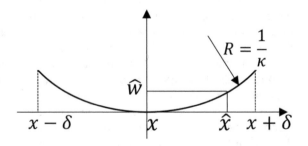

$$W_{PD} = \frac{1}{2}\alpha \left[\int_{x-\delta}^{x+\delta} \beta(|\hat{x} - x|)d\hat{x} \right]^2 \kappa^2. \tag{3.20}$$

Comparing above peridynamic beam deformation energy density function to the local Euler–Bernoulli beam deformation energy function at certain cross-section. The following relation is derived. That is

$$\alpha \left[\int_{x-\delta}^{x+\delta} \beta(|\hat{x} - x|)d\hat{x} \right]^2 = EI \tag{3.21}$$

where E is the Young's modulus of the material, I is the second moment of the beam cross-section. Further, to guarantee the consistency of nonlocal curvature to the local curvature for pure bending configuration as the horizon value δ goes to zero, the β must satisfy the following relations

$$\int_{x-\delta}^{x+\delta} \beta(|\hat{x} - x|)d\hat{x} = 1. \tag{3.22}$$

Then, the parameter α is further calibrated as

$$\alpha = EI. \tag{3.23}$$

Therefore, the peridynamic bending equation is

$$M = EI \int_{x-\delta}^{x+\delta} 2\beta(|\hat{x} - x|)\frac{\hat{w} - w}{(\hat{x} - x)^2}d\hat{x}. \tag{3.24}$$

Note that parameter β is function defined on the neighborhood of the point. It can be proved that by reducing the size of the horizon δ, the peridynamic bending Eq. (3.24) reduces to the classical Euler Bernoulli bending equation.

3.4 Peridynamic Differential Beam Equation

In the following derivation, the parameter β is assume to be constant over the horizon region for simplification. That is

$$\beta = \frac{1}{2\delta}. \tag{3.25}$$

Then, the reformulated peridynamic bending equation is

$$M = \frac{EI}{\delta} \int_{x-\delta}^{x+\delta} \frac{\hat{w} - w}{\left(\hat{x} - x\right)^2} dx'. \tag{3.26}$$

Note that, above bending equation is an integral form with nonlocal parameter δ. To compare this peridynamic bending equation to other differential form nonlocal bending equation, the beam deformation at neighboring point is reformulated by a Taylor's series expansion as

$$\hat{w} = w(x) + w^{(1)}(x) \cdot \left(\hat{x} - x\right) + \frac{1}{2!} w^{(2)}(x) \cdot \left(\hat{x} - x\right)^2$$
$$+ \frac{1}{3!} w^{(3)}(x) \cdot \left(\hat{x} - x\right)^3 + \cdots + \frac{1}{n!} w^{(M)}(x) \cdot \left(\hat{x} - x\right)^M + \cdots . \tag{3.27}$$

Note that above expression is expanded without considering pure bending deformation. Substitute above equation to the integral form peridynamic bending equation yields

$$M(x) = \frac{EI}{\delta} \int_{x-\delta}^{x+\delta} \left[\frac{w^{(1)}}{\hat{x} - x} + \frac{1}{2!} w^{(2)} + \frac{1}{3!} w^{(3)} \left(\hat{x} - x\right) + \frac{1}{4!} w^{(4)} \left(\hat{x} - x\right)^2 \right.$$
$$\left. + \cdots + \frac{1}{n!} w^{(n)} \left(\hat{x} - x\right)^{n-2} + \cdots \right] d\hat{x}. \tag{3.28}$$

That is,

$$M(x) = \sum_{n=1}^{\infty} a_n w^{(2n)}, \tag{3.29}$$

where

$$a_n = \frac{2EI\delta^{2n-2}}{(2n)!(2n-1)}, n = 1, 2, 3, 4, \ldots \tag{3.30}$$

Instead of deriving beam bending equation from nonlocal stress-strain relations, the above peridynamic beam equation derivation process starts from the definition of nonlocal curvature. The values of kernel functions α and β are calibrated from the consistencies of deformation energy and curvature for pure bending configuration. The integral form peridynamic bending Eqs. (3.24) and (3.26) can be simplified to the differential form bending Eq. (3.28) up to the infinite order. Therefore, by applying different boundary conditions, the analytical solution to the truncated order bending equation can be derived.

References

1. Silling S (2000) Reformulation of elasticity theory for discontinuities and long-range forces. J Mech Phys Solids 48:175–209
2. Silling S, Askari E (2005) A meshfree method based on the peridynamic model of solid mechanics. Comput Struct 83:1526–1535
3. Macek R, Silling S (2007) Peridynamic via finite element analysis. Finite Element Anal Des 43:1169–1178
4. Chen X, Gunzburger M (2011) Continuous and discontinuous finite element methods for a peridynamic model of mechanics. Comput Method Appl Mech Eng 200:1237–1250
5. Trageser J, Seleson P (2020) Bond-based peridynamics: a tale of two Poisson's ratios. J Peridyn Nonlocal Model 2:278–288
6. Silling S, Epton M, Weckner O (2007) Peridynamic states and constitutive modeling. J Elast 88:151–184
7. Moyer ET, Miraglia MJ (2014) Peridynamic solution for Timoshenko beams. Engineering 6:304–317
8. O'Grady J, Foster J (2014) Peridynamic beams: a non-ordinary, state-based model. Int J Solids Struct 51:3177–3183
9. O'Grady J, Foster J (2014) Peridynamic plates and flat shells: a non-ordinary, state-based model. Int J Solids Struct 51:4572–4579
10. O'Grady J, Foster J (2016) A meshfree method for bending and failure in non-ordinary peridynamic shells. Comput Mech 57:921–929
11. Diyaroglu C, Oterkus E, Oterkus S, Madenci E (2015) Peridynamics for bending of beams and plates with transverse shear deformation. Int J Solids Struct 69–70:152–168
12. Diyaroglu C, Oterkus E, Oterkus S (2019) An Euler–Bernoulli beam formulation in an ordinary state-based peridynamic framework. Math Mech Solids 24(2):361–376

Chapter 4
Analytical Solution to Benchmark Examples

This chapter analytically derives the deformation of beam governed by the Eringen's stress-gradient beam equation [1–4], the Mindlin's strain-gradient beam equation [5, 6], the higher-order beam equation [7] and the truncated peridynamic beam equation [8, 9]. The benchmark examples are solved including the simply-supported beam, the clamped–clamped beam and the cantilever beam. For comparison and simplification purpose, the cross-section of the beam is assumed to be constant over the length of the neutral axis. The distributed loads are assumed to be uniform.

4.1 Example 1: Simply-Supported Beam

The configuration of the simply-supported beam subjected to the uniformly distributed loads is illustrated in Fig. 4.1.

The value of the distributed load is assumed to be constant as

$$q(x) = Q. \tag{4.1}$$

Then, the bending moment over the beam can be derived after calculating the forces on the constraints. That is,

$$M = \frac{1}{2}Qx^2 - \frac{1}{2}QLx. \tag{4.2}$$

The boundary conditions are.

$x = 0$:

$$w(x)|_{x=0} = 0 \tag{4.3}$$

© The Author(s), under exclusive license to Springer Nature Switzerland AG 2021
J. Chen, *Nonlocal Euler–Bernoulli Beam Theories*,
SpringerBriefs in Continuum Mechanics,
https://doi.org/10.1007/978-3-030-69788-4_4

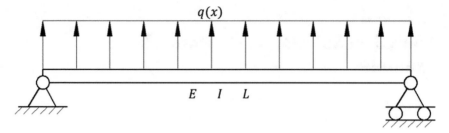

Fig. 4.1 Illustration of simply-supported beam. The Young's modulus, moment of inertia and length of the beam are E I and L

$$\left.\frac{\partial^2 w(x)}{\partial x^2}\right|_{x=0} = 0 \tag{4.4}$$

and $x = L$:

$$w(x)|_{x=L} = 0 \tag{4.5}$$

$$\left.\frac{\partial^2 w(x)}{\partial x^2}\right|_{x=L} = 0 \tag{4.6}$$

Note that Eqs. (4.3) and (4.5) are the 'physical' boundary conditions and is sufficient to solve the local Euler–Bernoulli bending equation which is the 2nd order differential equation on transverse displacement w. The 'physical' boundary condition means the transverse displacement at the hinge is zero. For the nonlocal beam bending equations whose governing equation is higher order differential equation on transverse displacement w, additional boundary conditions in Eqs. (4.4) and (4.6) required. Note that, the additional boundary conditions are built on the assumption that the bending moment on boundary calculated from the local Euler Bernoulli beam equation is zero. Considering the fact that the transverse displacement calibrated from the additional boundary conditions in Eqs. (4.4) and (4.6) satisfy the zero bending moment condition for both local bending equation and nonlocal bending equation, we call these additional constraints as 'local-nonlocal bending moment consistency constraint'. Note that the expressions of these additional boundary conditions are not unique. For example, the addition boundary conditions can be defined based on the shear force local-nonlocal consistency condition. That is, the transverse displacement calculated from the nonlocal bending equation has to satisfy the local shear force expression on boundaries. For simplification, this subsection utilizes the boundary conditions in Eqs. (4.3)–(4.6).

The transverse displacement of the simply-supported beam configured in Fig. 4.1 can be immediately calculated from local Euler–Bernoulli bending moment equation. That is,

$$w(x) = \frac{1}{24EI}QL^3x - \frac{1}{12EI}QLx^3 + \frac{1}{24EI}Qx^4. \tag{4.7}$$

The following subsection will calculate the beam transverse displacement subjected to above loading and boundary conditions governed by different nonlocal beam equations respectively.

4.1.1   Eringen's Stress-Gradient Beam Equation

The stress–strain relation in Eringen's stress-gradient beam model exerts differential operation on the stress term, which results in additional second order differential term of bending moment on the beam equation. Considering the fact that the differential term of the transverse displacement is still second order, the Eringen's stress-gradient beam equation can be directly solved by integrating on both sides of the Eq. (2.5) after substituting bending moment expression in Eq. (4.2). That is

$$w = c_1 + c_2 x + \frac{1}{24EI}Qx^4 - \frac{1}{12EI}QLx^3 - \frac{\delta_{eg}^2}{2EI}Qx^2, \tag{4.8}$$

where c_1 and c_2 are constants and can be calibrated from the boundary condition in Eqs. (4.3) and (4.5). Those are

$$c_1 = 0 \tag{4.9}$$

$$c_2 = \frac{1}{24EI}QL^3 + \frac{\delta_{eg}^2}{2EI}QL. \tag{4.10}$$

Note that, the physical boundary conditions in Eqs. (4.3) and (4.5) are sufficient to calculate the parameters c_1 and c_2. In other words, the calibrated transverse displacement fails to satisfy the bending moment consistency constraints in Eqs. (4.4) and (4.6) in general. This can be verified by substituting the following transverse displacement into Eqs. (4.4) and (4.6).

$$w = \left(\frac{1}{24EI}QL^3 + \frac{\delta_{eg}^2}{2EI}QL \right)x + \frac{1}{24EI}Qx^4 - \frac{1}{12EI}QLx^3 - \frac{\delta_{eg}^2}{2EI}Qx^2. \tag{4.11}$$

The transverse displacements at different nonlocal parameter are shown in Fig. 4.2a. It can be observed that nonlocal transverse displacement is close to the local solution when the nonlocal parameter δ_{eg} is small. As a result of the nonlocal effect, the displacement of the nonlocal beam differs from the local beam deformation with the increase of the nonlocal parameter δ_{eg}. And the nonlocal effect governed by

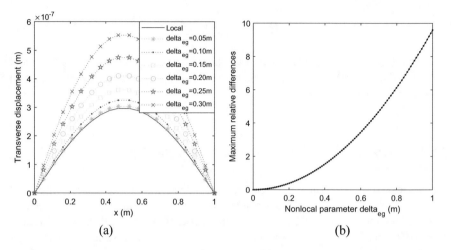

Fig. 4.2 Transverse displacements of the simply-supported beam governed by Eringen's stress-gradient beam theory. **a** the transverse displacements over the beam; **b** the relative nonlocal-local difference w.r.t. nonlocal parameter δ_{eg}

Eringen's stress-gradient beam equation makes the beam 'softer'. Moreover, the relative difference between nonlocal solutions to the local solution at different nonlocal parameters is shown in Fig. 4.2b. It can be observed that relative nonlocal-local difference accumulated rapidly with the increase of the nonlocal parameter.

4.1.2   Mindlin's Strain-Gradient Beam Equation

The Mindlin's strain-gradient bending equation is described in Eq. (2.7). To solve the strain-gradient bending equation which involves the 2nd and the 4th order differential terms, the polynomial characteristic equation corresponding to the beam equation is introduced. That is

$$\lambda^2 - \delta_{sg}{}^2 \lambda^4 = 0. \tag{4.12}$$

Solve above equation yields

$$\lambda_1 = 0; \; \lambda_2 = 0; \; \lambda_3 = \frac{1}{\delta_{sg}}; \; \lambda_4 = -\frac{1}{\delta_{sg}};$$

Then, the complimentary solution is expanded as

$$w_c(x) = c_1 + c_2 x + c_3 e^{\frac{1}{\delta_{sg}}x} + c_4 e^{-\frac{1}{\delta_{sg}}x} \tag{4.13}$$

Substitute the bending moment Eq. (4.2) into Eq. (2.7), the strain-gradient beam equation is rewritten as

$$w^{(2)} - \delta_{sg}^{2} w^{(4)} = \frac{1}{2EI} Q x^{2} - \frac{1}{2EI} Q L x \tag{4.14}$$

Considering the fact that the right-hand side of the Eq. (4.14) is a polynomial, one of the particular solutions to above differential equation (4.14) can be formulated as

$$w_p = a_0 + a_1 x + a_2 x^2 + a_3 x^3 + a_4 x^4 \tag{4.15}$$

where a_0, a_1, a_2, a_4, a_4 are constants and are calibrated after substituting Eq. (4.15) in to Eq. (4.14) and equalizing coefficients of the same order terms. Those are

$$a_2 = \frac{1}{2EI} Q \delta_{sg}^{2} \tag{4.16}$$

$$a_3 = -\frac{1}{12EI} QL \tag{4.17}$$

$$a_4 = \frac{1}{24EI} Q. \tag{4.18}$$

Then, the solution to the Mindlin's strain-gradient beam equation can be formulated by combining the complimentary solution and one of the particular solutions. That is

$$w = c_1 + c_2 x + c_3 e^{\frac{1}{\delta_{sg}} x} + c_4 e^{-\frac{1}{\delta_{sg}} x} + \frac{1}{2EI} Q \delta_{sg}^{2} x^2$$
$$- \frac{1}{12EI} Q L x^3 + \frac{1}{24EI} Q x^4 \tag{4.19}$$

Note that c_1, c_2, c_3 and c_4 are constants and are required to be calibrated from the nonlocal boundary conditions. After substituting Eq. (4.19) into Eqs. (4.3) to (4.6), the coefficients are calibrated. Those are

$$c_1 = \frac{1}{EI} Q \delta_{sg}^{4} \tag{4.20}$$

$$c_2 = \frac{1}{24EI} Q L^3 - \frac{1}{2EI} Q \delta_{sg}^{2} L \tag{4.21}$$

$$c_3 = -\frac{1}{EI} Q \delta_{sg}^{4} \frac{1}{\left(e^{\frac{1}{\delta_{sg}} L} + 1 \right)} \tag{4.22}$$

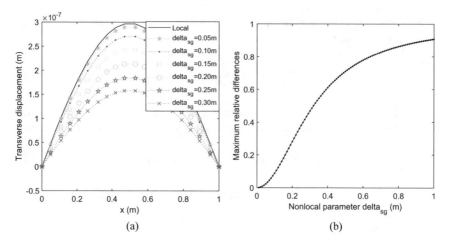

Fig. 4.3 Transverse displacements of the simply-supported beam governed by Mindlin's strain-gradient beam. **a** the transverse displacements over the beam; **b** the relative nonlocal-local difference w.r.t. nonlocal parameter δ_{sg}

$$c_4 = -\frac{1}{EI} Q\delta_{sg}{}^4 \frac{e^{\frac{1}{\delta_{sg}}L}}{\left(e^{\frac{1}{\delta_{sg}}L} + 1\right)}. \tag{4.23}$$

The coefficients of the transverse displacement in Eq. (4.19) are calibrated from Eqs. (4.3) to (4.6), thus it satisfies both the physical boundary conditions and the local-nonlocal bending moment constraints. The transverse displacements of Mindlin's strain-gradient beam at different nonlocal parameter are shown in Fig. 4.3a. The relative difference between nonlocal solutions to the local solution is shown in Fig. 4.3b. It can be observed that the nonlocal effect appears with the increase of nonlocal parameter δ_{sg}. And the nonlocal effect governed by the Mindlin's strain-gradient beam equation makes the beam much 'stiffer'.

4.1.3 Higher-Order Beam Equation

Considering the fact that the characteristic equation of higher-order beam Eq. (2.9) is identical to the strain-gradient beam equation. The complimentary solution to Eq. (2.9) has the same form as Eq. (4.13). After substituting the bending moment into Eq. (2.9), the higher-order beam equation is rewritten as

$$w^{(2)} - \delta_{sg}{}^2 w^{(4)} = \frac{1}{2EI} Qx^2 - \frac{1}{2EI} QLx - \frac{\delta_{eg}{}^2}{EI} Q. \tag{4.24}$$

By assuming the particular solution be polynomial as Eq. (4.15), the coefficient of the particular solution can be calibrated by equalizing the coefficients of terms with same order. Then, the particular solution can be formulated as

$$w_p = \frac{Q}{2EI}\left(\delta_{sg}^2 - \delta_{eg}^2\right)x^2 - \frac{1}{12EI}QLx^3 + \frac{1}{24EI}Qx^4. \tag{4.25}$$

After combining the complimentary solution in Eq. (4.13) and the calibrated particular solution in Eq. (4.25), the solution to the higher-order beam equation is

$$w = c_1 + c_2 x + c_3 e^{\frac{1}{\delta_{sg}}x} + c_4 e^{-\frac{1}{\delta_{sg}}x} + \frac{Q}{2EI}\left(\delta_{sg}^2 - \delta_{eg}^2\right)x^2$$
$$- \frac{1}{12EI}QLx^3 + \frac{1}{24EI}Qx^4 \tag{4.26}$$

where c_1, c_2, c_3 and c_4 are constants and are required to be calibrated from the boundary conditions. After substituting Eq. (4.26) into Eqs. (4.3) to (4.6), the coefficients are calibrated. Those are

$$c_1 = \frac{Q\delta_{sg}^2}{EI}\left(\delta_{sg}^2 - \delta_{eg}^2\right) \tag{4.27}$$

$$c_2 = \frac{1}{24EI}QL^3 - \frac{Q}{2EI}\left(\delta_{sg}^2 - \delta_{eg}^2\right)L \tag{4.28}$$

$$c_3 = -\frac{Q\delta_{sg}^2}{EI}\left(\delta_{sg}^2 - \delta_{eg}^2\right)\frac{1}{\left(e^{\frac{1}{\delta_{sg}}L} + 1\right)} \tag{4.29}$$

$$c_4 = -\frac{Q\delta_{sg}^2}{EI}\left(\delta_{sg}^2 - \delta_{eg}^2\right)\frac{e^{\frac{1}{\delta_{sg}}L}}{\left(e^{\frac{1}{\delta_{sg}}L} + 1\right)} \tag{4.30}$$

Note that the higher-order beam equation involves two nonlocal parameters. By reducing one of the nonlocal parameters to zero, the higher-order beam equation and its transverse displacement are reduced to the Eringen's stress-gradient equation or the Mindlin's strain-gradient equation.

The influence of these two nonlocal parameters are investigated by fixing the value of one nonlocal parameter and changing the value of the other parameter. The transverse displacement of the higher-order beam at different values of nonlocal parameter δ_{eg} after fixing δ_{sg} are show in Fig. 4.4. When the nonlocal parameter δ_{sg} is small, the deformation curve with $\delta_{eg} = 0.05$ is closest to the local solution as shown in Fig. 5a. For a larger value of the $\delta_{sg} = 0.18$, the solutions with nonlocal parameter δ_{eg} equals to 0.15 and 0.20 are much closer to the local solution. Similar conclusion can be observed by varying the nonlocal parameter δ_{sg} while fixing δ_{eg}, as shown in Fig. 4.5. It can further be observed that different from the stress-gradient

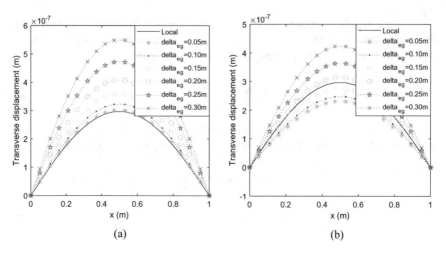

Fig. 4.4 Transverse displacements of the simply-supported beam governed by higher-order beam. **a** the transverse displacements at $\delta_{sg} = 0.03$; **b** the transverse displacements at $\delta_{sg} = 0.18$

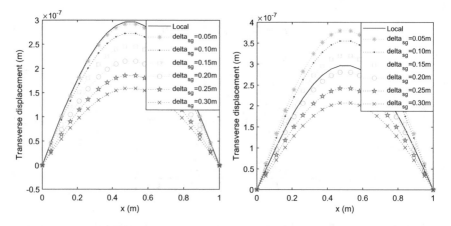

Fig. 4.5 Transverse displacements of the simply-supported beam governed by higher-order beam. **a** the transverse displacements at $\delta_{eg} = 0.03$; **b** the transverse displacements at $\delta_{eg} = 0.18$

and the strain-gradient beam models, the higher-order beam equation can be much stiffer or softer to the local beam by varying the values of the nonlocal parameters. In details, the nonlocal parameter δ_{eg} tends to make the beam 'softer' while the nonlocal parameter δ_{sg} makes the beam 'stiffer'. The relative nonlocal-local difference at different nonlocal parameters are shown in Fig. 4.6. It can be observed that the higher-order beam solution is close to the local beam solution as the nonlocal parameters are close to each other.

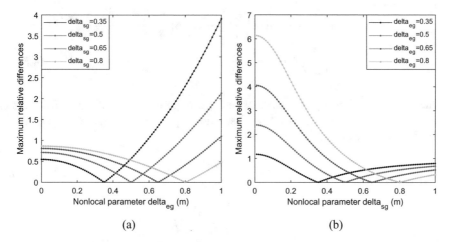

Fig. 4.6 Relative nonlocal-local displacement differences of the simply-supported beam to the nonlocal parameter

4.1.4 Truncated Peridynamic Beam Equation

The peridynamic integral-form beam equation can be expanded in to differential equation up to infinite order by Taylor series expansion. In this sub-section, the differential-form peridynamic beam equation is truncated up to the 4th order. That is

$$w_x^{(2)} + \frac{\delta^2}{36} w_x^{(4)} = \frac{M(x)}{EI}. \tag{4.31}$$

After solving the polynomial characteristic equation, the complimentary solution to above 4th order differential equation can be written as

$$w_c(x) = c_1 + c_2 x + c_3 \cos\left(\frac{6}{\delta}x\right) + c_4 \sin\left(\frac{6}{\delta}x\right). \tag{4.32}$$

Following the same procedures on calculating the particular solution of the nonlocal bending equations, the particular solution of the 4th order peridynamic bending equation is formulated. That is

$$w_p = -\frac{\delta^2}{72EI} Qx^2 - \frac{1}{12EI} QLx^3 + \frac{1}{24EI} Qx^4. \tag{4.33}$$

Then, the full solution to the truncated 4th order peridynamic beam equation is

$$w(x) = c_1 + c_2 x + c_3 \cos\left(\frac{6}{\delta}x\right) + c_4 \sin\left(\frac{6}{\delta}x\right) - \frac{\delta^2}{72EI}Qx^2$$
$$- \frac{1}{12EI}QLx^3 + \frac{1}{24EI}Qx^4 \tag{4.34}$$

where c_1, c_2, c_3 and c_4 are constants and are required to be calibrated from the nonlocal boundary conditions. After substituting Eq. (4.34) into Eqs. (4.3)–(4.6), the coefficients are calculated. Those are

$$c_1 = \frac{Q\delta^4}{1296EI} \tag{4.35}$$

$$c_2 = \frac{QL^3}{24EI} + \frac{QL\delta^2}{72EI} \tag{4.36}$$

$$c_3 = -\frac{Q\delta^4}{1296EI} \tag{4.37}$$

$$c_4 = \frac{Q\delta^4}{1296EI}\left(\frac{\cos\left(\frac{6}{\delta}L\right)}{\sin\left(\frac{6}{\delta}L\right)} - \frac{1}{\sin\left(\frac{6}{\delta}L\right)}\right) \tag{4.38}$$

The transverse displacement of the Euler–Bernoulli beam governed by the truncated 4th order peridynamic bending equation is shown in Fig. 4.7a. The nonlocal-local difference, which is defined as the nonlocal solutions minus the local solution, is shown in Fig. 4.70b. It can be observed that small nonlocal parameter δ generates a solution much closer to the local solution. Nonlocal behavior becomes much severe

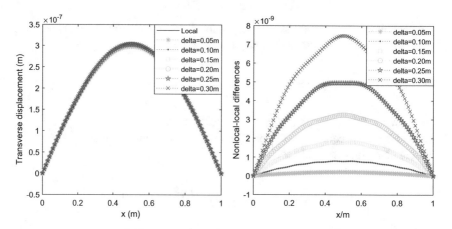

Fig. 4.7 Transverse displacement of the simply-supported beam governed by the truncated 4th order peridynamic beam. **a** Transverse displacement at different nonlocal parameters, **b** nonlocal-local differences at different nonlocal parameters

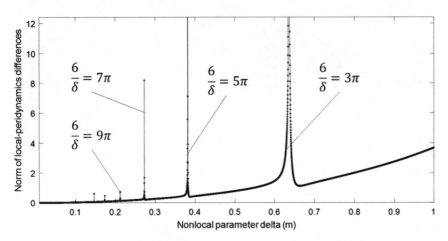

Fig. 4.8 Norm of nonlocal-local differences at different values of nonlocal parameter

for larger values of δ. Moreover, a much closer observation on Fig. 4.70b shows that a slight fluctuation appears on the nonlocal deformations, for example the deformation at $\delta = 0.25$ and $\delta = 0.3$.

To further investigate the influence of nonlocal parameter on the response of the truncated 4th order perdynamic beam, the norm of the nonlocal-local differences is calculated at different values of nonlocal parameter, as shown in Fig. 4.8. Different from the Eringen's stress-gradient and Mindlin's strain-gradient nonlocal beam theories, in which the nonlocal-local difference changes asymptotically as a function of the nonlocal parameter, the nonlocal-local difference of the truncated 4th order peridynamic beam equation increases rapidly at certain values of the nonlocal parameter. That is, although the overall trend tends to increase with the increase of the nonlocal parameter as a result of the nonlocal effect, pulse exist on a series of certain values. To help explain this phenomenon, the transverse displacement in Eqs. (4.34)–(4.38) are investigated. Equation (4.38) involves $\sin\left(\frac{6}{\delta}L\right)$ term on the denominator, which introduces singularity problem $\frac{6}{\delta}L$ equals odd number times of π. To further investigates the influence of the singularity, the beam responses at nonlocal parameter closes to singular values are investigated, as shown in Fig. 4.9. Figure 4.9 shows that nonlocal parameter close to singular value $\delta = 0.27$ generates largest fluctuation on the transverse displacement. Figure 4.9b indicates the singular value of the nonlocal parameter is close to 0.38 as the transverse fluctuation is the largest at this value. These results on Fig. 4.9 are consistent to Fig. 4.8.

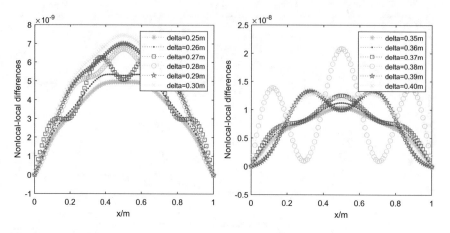

Fig. 4.9 Transverse displacement of the simply-supported beam at nonlocal parameter close to singular value. **a** Nonlocal parameters near singular value 0.27 m, **b** nonlocal parameters near singular value 0.38 m

4.2 Example 2: Clamped–Clamped Beam

The clamped–clamped beam subjected to distributed loads is illustrated in Fig. 4.10.

The value of distributed load is assumed to be constant Q. The boundary conditions are.

$x = 0$:

$$w(x)|_{x=0} = 0 \tag{4.39}$$

$$\frac{\partial w(x)}{\partial x}\bigg|_{x=0} = 0 \tag{4.40}$$

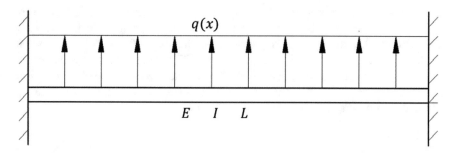

Fig. 4.10 Illustration of clamped–clamped beam. The Young's modulus, moment of inertia and length of the beam are E I and L

and $x = L$:

$$w(x)|_{x=L} = 0 \tag{4.41}$$

$$\frac{\partial w(x)}{\partial x}\bigg|_{x=L} = 0 \tag{4.42}$$

Then, the bending moment over the beam can be derived after calculating the forces and the moments on the constraints and considering the deformation compliance condition on boundary. That is,

$$M = \frac{1}{2}Qx^2 - \frac{1}{2}QLx + \frac{1}{12}QL^2. \tag{4.43}$$

Note that, in this clamped–clamped beam configuration, the boundary conditions in Eqs. (4.39)–(4.42) are all physical boundary condition. That is, the transverse displacement and the slop on the clamped end are zero. The local-nonlocal consistency condition is utilized for calculating the concentrated moments on the boundary from the deformation compliance condition. That is, the bending moments on the boundaries calculated from both local and nonlocal bending equations are equal. The deformation of the clamped–clamped beam in Fig. 4.10 can be immediately calculated from local Euler–Bernoulli bending moment equation. That is,

$$w(x) = -\frac{QL^2}{24EI}x^2 + \frac{QL}{12EI}x^3 - \frac{Q}{24EI}x^4. \tag{4.44}$$

The following subsection will calculate the beam transverse displacement subjected to above loading and boundary conditions governed by different nonlocal beam theories respectively. The calculation procedures are identical to the simply-supported beam except for the bending moment expression and the boundary conditions.

4.2.1 Eringen's Stress-Gradient Beam Equation

The transverse deformation of the beam governed by Eringen's stress-gradient beam equation is calculated as

$$\begin{aligned}
w = {} & \frac{1}{24EI}Qx^4 - \frac{1}{12EI}QLx^3 + \frac{1}{24EI}QL^2x^2 \\
& - \frac{1}{2EI}Q\delta_{eg}^2x^2 + \frac{Q\delta_{eg}^2L}{2EI}x
\end{aligned} \tag{4.45}$$

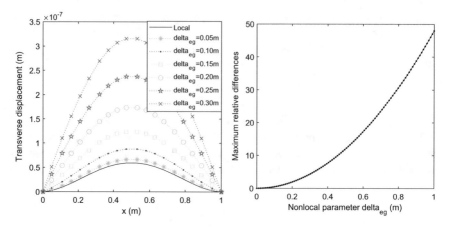

Fig. 4.11 Transverse displacements of clamped–clamped beam governed by Eringen's stress-gradient beam theory. **a** the transverse displacements over the beam; **b** the relative nonlocal-local difference w.r.t. nonlocal parameter δ_{eg}

The transverse displacements of clamped–clamped Euler–Bernoulli beam governed by Eringen's stress-gradient equation at different nonlocal parameter are shown in Fig. 4.11a. Similar to the deformation of the simply-supported beam, the nonlocal transverse displacement of the clamped–clamped beam is close to the local solution when the nonlocal parameter δ_{eg} is small. And the displacement of the nonlocal beam differs from the local beam deformation with the increase of the nonlocal parameter δ_{eg}. The relative difference between nonlocal solutions to the local solution at different nonlocal parameters in Fig. 4.11b confirms the same conclusion that relative nonlocal-local difference or the nonlocal effect accumulated rapidly with the increase of the nonlocal parameter.

4.2.2 Mindlin's Strain-Gradient Beam Equation

The transverse deformation of the beam governed by Mindlin's strain-gradient beam equation is calculated as

$$w = c_1 + c_2 x + c_3 e^{\frac{1}{\delta_{sg}}x} + c_4 e^{-\frac{1}{\delta_{sg}}x} + \left(\frac{1}{24EI}QL^2 + \frac{1}{2EI}Q\delta_{sg}^2\right)x^2$$
$$- \frac{1}{12EI}QLx^3 + \frac{1}{24EI}Qx^4 \tag{4.46}$$

where c_1, c_2, c_3 and c_4 are constants and are required to be calibrated from the nonlocal boundary conditions. After substituting Eq. (4.46) into Eqs. (4.39)–(4.42), the coefficients are calculated. Those are

$$c_1 = \frac{Q\delta_{sg}{}^3 L\left[L\left(e^{-\frac{1}{\delta_{sg}}L} + e^{\frac{1}{\delta_{sg}}L}\right) + 2\delta_{sg}\left(e^{-\frac{1}{\delta_{sg}}L} - e^{\frac{1}{\delta_{sg}}L}\right) + 2L\right]}{8EI\delta_{sg} - 4EI\delta_{sg}\left(e^{\frac{1}{\delta_{sg}}L} + e^{-\frac{1}{\delta_{sg}}L}\right) + 2EIL\left(e^{\frac{1}{\delta_{sg}}L} - e^{-\frac{1}{\delta_{sg}}L}\right)} \tag{4.47}$$

$$c_2 = \frac{Q\delta_{sg}{}^2 L\left[L\left(e^{-\frac{1}{\delta_{sg}}L} - e^{\frac{1}{\delta_{sg}}L}\right) + 2\delta_{sg}\left(e^{-\frac{1}{\delta_{sg}}L} + e^{\frac{1}{\delta_{sg}}L}\right) - 4\delta_{sg}\right]}{8EI\delta_{sg} - 4EI\delta_{sg}\left(e^{\frac{1}{\delta_{sg}}L} + e^{-\frac{1}{\delta_{sg}}L}\right) + 2EIL\left(e^{\frac{1}{\delta_{sg}}L} - e^{-\frac{1}{\delta_{sg}}L}\right)} \tag{4.48}$$

$$c_3 = \frac{-Q\delta_{sg}{}^3 L\left(Le^{-\frac{1}{\delta_{sg}}L} + 2\delta_{sg}e^{-\frac{1}{\delta_{sg}}L} + L - 2\delta_{sg}\right)}{8EI\delta_{sg} - 4EI\delta_{sg}\left(e^{\frac{1}{\delta_{sg}}L} + e^{-\frac{1}{\delta_{sg}}L}\right) + 2EIL\left(e^{\frac{1}{\delta_{sg}}L} - e^{-\frac{1}{\delta_{sg}}L}\right)} \tag{4.49}$$

$$c_4 = \frac{-Q\delta_{sg}{}^3 L\left(Le^{\frac{1}{\delta_{sg}}L} - 2\delta_{sg}e^{\frac{1}{\delta_{sg}}L} + L + 2\delta_{sg}\right)}{8EI\delta_{sg} - 4EI\delta_{sg}\left(e^{\frac{1}{\delta_{sg}}L} + e^{-\frac{1}{\delta_{sg}}L}\right) + 2EIL\left(e^{\frac{1}{\delta_{sg}}L} - e^{-\frac{1}{\delta_{sg}}L}\right)} \tag{4.50}$$

The transverse displacements of the clamped–clamped beam governed by Mindlin's strain-gradient beam at different nonlocal parameter are shown in Fig. 4.12a. The relative difference between nonlocal solutions to the local solution is shown in Fig. 4.12b. Similar to the observation in Fig. 4.3, the nonlocal effect increases with the increase of nonlocal parameter δ_{sg}. And the nonlocal effect governed by the Mindlin's strain-gradient beam equation makes the beam much 'stiffer'.

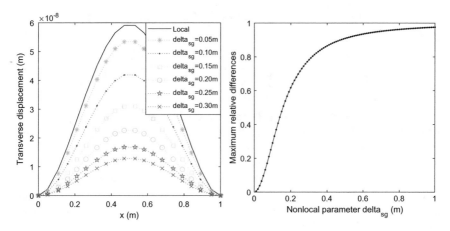

Fig. 4.12 Transverse displacements of clamped–clamped beam governed by Mindlin's strain-gradient beam theory at different nonlocal parameters. **a** the transverse displacements over the beam; **b** the relative nonlocal-local difference w.r.t. nonlocal parameter δ_{eg}

4.2.3 Higher-Order Beam Equation

The transverse deformation of the beam governed by higher-order beam equation is calculated as

$$w = c_1 + c_2 x + c_3 e^{\frac{1}{\delta_{sg}}x} + c_4 e^{-\frac{1}{\delta_{sg}}x} + \left[\frac{1}{24EI}QL^2 + \frac{1}{2EI}Q\left(\delta_{sg}^2 - \delta_{eg}^2\right)\right]x^2$$
$$- \frac{1}{12EI}QLx^3 + \frac{1}{24EI}Qx^4 \tag{4.51}$$

where c_1, c_2, c_3 and c_4 are constants and are required to be calibrated from the nonlocal boundary conditions. After substituting Eq. (4.51) into Eqs. (4.39)–(4.42), the coefficients are calculated. Those are

$$c_1 = -\frac{QL\delta_{sg}\left(\delta_{sg}^2 - \delta_{eg}^2\right)\left[L\left(e^{-\frac{1}{\delta_{sg}}L} + e^{\frac{1}{\delta_{sg}}L}\right) + 2\delta_{sg}\left(e^{-\frac{1}{\delta_{sg}}L} - e^{\frac{1}{\delta_{sg}}L}\right) + 2L\right]}{2EIL\left(e^{-\frac{1}{\delta_{sg}}L} - e^{\frac{1}{\delta_{sg}}L}\right) + 4EI\delta_{sg}\left(e^{-\frac{1}{\delta_{sg}}L} + e^{\frac{1}{\delta_{sg}}L}\right) - 8EI\delta_{sg}} \tag{4.52}$$

$$c_2 = -\frac{QL\left(\delta_{sg}^2 - \delta_{eg}^2\right)}{2EI} \tag{4.53}$$

$$c_3 = \frac{QL\delta_{sg}\left(\delta_{sg}^2 - \delta_{eg}^2\right)\left(Le^{-\frac{1}{\delta_{sg}}L} + L - 2\delta_{sg} + 2\delta_{sg}e^{-\frac{1}{\delta_{sg}}L}\right)}{2EIL\left(e^{-\frac{1}{\delta_{sg}}L} - e^{\frac{1}{\delta_{sg}}L}\right) + 4EI\delta_{sg}\left(e^{-\frac{1}{\delta_{sg}}L} + e^{\frac{1}{\delta_{sg}}L}\right) - 8EI\delta_{sg}} \tag{4.54}$$

$$c_4 = \frac{QL\delta_{sg}\left(\delta_{sg}^2 - \delta_{eg}^2\right)\left(Le^{\frac{1}{\delta_{sg}}L} + 2\delta_{sg} + L - 2\delta_{sg}e^{\frac{1}{\delta_{sg}}L}\right)}{2EIL\left(e^{-\frac{1}{\delta_{sg}}L} - e^{\frac{1}{\delta_{sg}}L}\right) + 4EI\delta_{sg}\left(e^{-\frac{1}{\delta_{sg}}L} + e^{\frac{1}{\delta_{sg}}L}\right) - 8EI\delta_{sg}} \tag{4.55}$$

Similar to the simply-supported beam by the higher-order beam equation, the influence of the two nonlocal parameters δ_{eg} and δ_{sg} are investigated respectively by fixing the value of one nonlocal parameter and changing the value of the other parameter, as shown in Figs. 4.13 and 4.14. It can be observed that the clamped–clamped beam governed by the higher-order beam equation can be much stiffer or softer to the local beam by varying the values of the nonlocal parameters. In details, the nonlocal parameter δ_{eg} tends to make the beam 'softer' while the nonlocal parameter δ_{sg} makes the beam 'stiffer'. The relative nonlocal-local difference at different nonlocal parameters are shown in Fig. 4.15. It can be observed that the higher-order beam solution is close to the local beam solution as the nonlocal parameters are close to each other.

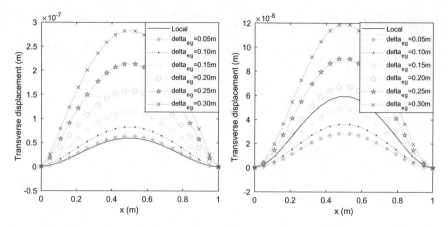

Fig. 4.13 Transverse displacements of the clamped–clamped beam governed by higher-order beam. **a** the transverse displacements at $\delta_{sg} = 0.03$; **b** the transverse displacements at $\delta_{sg} = 0.18$

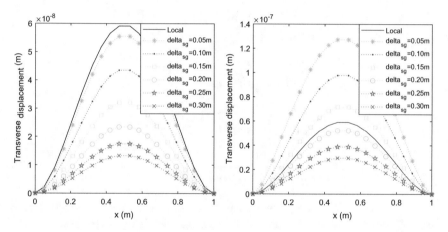

Fig. 4.14 Transverse displacements of the clamped–clamped beam governed by higher-order beam. **a** the transverse displacements at $\delta_{eg} = 0.03$; **b** the transverse displacements at $\delta_{eg} = 0.18$

4.2.4   Truncated Peridynamic Beam Equation

$$w = c_1 + c_2 x + c_3 \cos\left(\frac{6}{\delta}x\right) + c_4 \sin\left(\frac{6}{\delta}x\right) + \left(\frac{1}{24EI}QL^2 - \frac{1}{72EI}Q\delta^2\right)x^2$$

$$- \frac{1}{12EI}QLx^3 + \frac{1}{24EI}Qx^4 \tag{4.56}$$

where c_1, c_2, c_3 and c_4 are constants and are required to be calibrated from the nonlocal boundary conditions. After substituting Eq. (4.56) into Eqs. (4.39)–(4.42), the coefficients are calculated. Those are

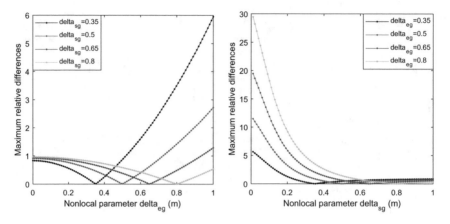

Fig. 4.15 Relative nonlocal-local displacement differences of the clamped–clamped beam to the nonlocal parameter

$$c_1 = \frac{Q\delta^3 L\left[3L + 3L\cos\left(\frac{6}{\delta}L\right) - \delta\sin\left(\frac{6}{\delta}L\right)\right]}{216EI\left[6L\sin\left(\frac{6}{\delta}L\right) + 2\delta\cos\left(\frac{6}{\delta}L\right) - 2\delta\right]} \tag{4.57}$$

$$c_2 = \frac{3Q\delta^2 L^2\sin\left(\frac{6}{\delta}L\right) + Q\delta^3 L\cos\left(\frac{6}{\delta}L\right) - Q\delta^3 L}{216EILsin\left(\frac{6}{\delta}L\right) + 72EI\delta\cos\left(\frac{6}{\delta}L\right) - 72EI\delta} \tag{4.58}$$

$$c_3 = -\frac{Q\delta^3 L\left[3L + 3L\cos\left(\frac{6}{\delta}L\right) - \delta\sin\left(\frac{6}{\delta}L\right)\right]}{216EI\left[6L\sin\left(\frac{6}{\delta}L\right) + 2\delta\cos\left(\frac{6}{\delta}L\right) - 2\delta\right]} \tag{4.59}$$

$$c_4 = -\frac{3Q\delta^3 L^2\sin\left(\frac{6}{\delta}L\right) + Q\delta^4 L\cos\left(\frac{6}{\delta}L\right) - Q\delta^4 L}{1296EILsin\left(\frac{6}{\delta}L\right) + 432EI\delta\cos\left(\frac{6}{\delta}L\right) - 432EI\delta} \tag{4.60}$$

The transverse displacement of the clamped–clamped beam governed by the truncated 4th order peridynamic bending equation is shown in Fig. 4.16. Similar to the simply-supported beam, two conclusions can be observed. First, the increase of the nonlocal parameter δ increases the nonlocal effect of the beam, which makes the beam much softer. Second, small fluctuation of the transverse displacement still exists at particular values of nonlocal parameter δ. Similar to the explanation of the fluctuation on simply-supported beam, this fluctuation results from the singularity of the expression on the transverse displacement.

4.3   Example 3: Cantilever Beam

The configuration of the simply-supported beam subjected to the distributed loads is illustrated in Fig. 4.17.

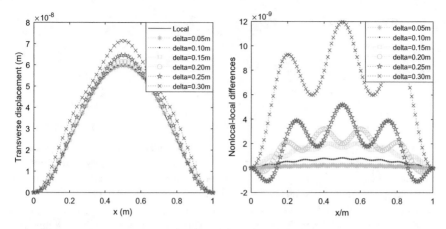

Fig. 4.16 Transverse displacement of the clamped–clamped beam governed by the truncated 4th order peridynamic beam. **a** Transverse displacement at different nonlocal parameters, **b** nonlocal-local differences at different nonlocal parameters

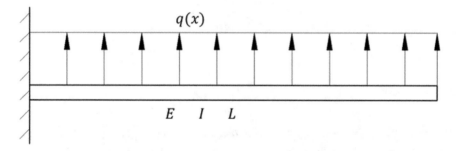

Fig. 4.17 Illustration of cantilever beam. The Young's modulus, moment of inertia and length of the beam are E I and L

By assuming the constant cross-section and the uniform distribution load $q(x) = Q$, the bending moment along the cantilever beam is

$$M = \frac{1}{2}Q(x - L)^2. \tag{4.61}$$

The boundary conditions are defined at the clamped end. That is, $x = 0$:

$$w(x)|_{x=0} = 0, \tag{4.62}$$

$$\frac{\partial w(x)}{\partial x}\bigg|_{x=0} = 0, \tag{4.63}$$

$$\frac{\partial^2 w(x)}{\partial x^2}\bigg|_{x=0} = \frac{QL^2}{2EI}, \tag{4.64}$$

and

$$\frac{\partial^3 w(x)}{\partial x^3}\bigg|_{x=0} = -\frac{QL}{EI}. \tag{4.65}$$

Similar to the boundary conditions of the simply-supported beam, the boundary conditions of the nonlocal cantilever beam are categorized into the physical boundary condition in Eqs. (4.62) and (4.63) and the local-nonlocal consistency conditions of the bending moment (Eq. 4.64) and the shear force (Eq. 4.65). That is, the transverse displacement calculated from the following nonlocal beam theories also satisfies the local constraints of classical Euler Bernoulli beam at the clamped end. Note that, the local-nonlocal consistency conditions can also be defined on the free end. That is,

$x = L$:

$$\frac{\partial^2 w(x)}{\partial x^2}\bigg|_{x=L} = 0, \tag{4.66}$$

and

$$\frac{\partial^3 w(x)}{\partial x^3}\bigg|_{x=L} = 0. \tag{4.67}$$

Then, Eqs. (4.62, 4.63, 4.66, 4.67) define a nonlocal boundary conditions, which are sufficient to derive the transverse displacement of the beam [10, 11]. The transverse deformation derived from the classical Euler–Bernoulli beam equation can be calculated by utilizing boundary condition in Eqs. (4.62) and (4.63). That is,

$$w(x) = \frac{QL^2}{4EI}x^2 - \frac{QL}{6EI}x^3 + \frac{Q}{24EI}x^4. \tag{4.68}$$

4.3.1   Eringen's Stress-Gradient Beam Equation

The general solution to the Eringen's stress-gradient beam bending equation is formulated after substituting the bending moment distribution Eq. (4.61). That is,

$$w = c_1 + c_2 x + \frac{1}{24EI}Qx^4 - \frac{QL}{6EI}x^3 + \frac{1}{4EI}QL^2x^2 - \frac{\delta_{eg}^2}{2EI}Qx^2 \tag{4.69}$$

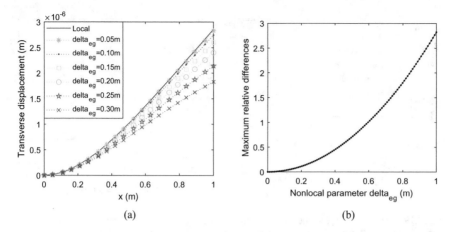

Fig. 4.18 Transverse displacement of the cantilever beam governed by the Eringen's stress gradient beam equation. **a** Transverse displacement at different nonlocal parameters, **b** Maximum nonlocal-local differences

The general transverse displacement expression contains two unknows which can be calibrated from the two constraints on the clamped end. The physical boundary conditions in Eqs. (4.62) (4.63) are utilized for calibration. Then, the calibrated transverse displacement is

$$w = \frac{1}{24EI}Qx^4 - \frac{QL}{6EI}x^3 + \left(\frac{QL^2}{4EI} - \frac{Q\delta_{eg}^2}{2EI}\right)x^2 \qquad (4.70)$$

Note that above Eq. (4.70) reduces to Eq. (4.68) when nonlocal parameter δ_{eg} goes to zero. Also, the nonlocal transverse displacement fails to satisfy the local-nonlocal consistency constraints in general. The transverse displacement at different nonlocal parameter δ_{eg} is shown in Fig. 4.18. It can be observed that the cantilever beam governed by Eringen's stress-gradient beam equation is much stiffer. This is different from the observations shown in Figs. 4.2 (4.11). Moreover, the transverse displacement of this nonlocal beam reduces to the local Euler Bernoulli beam as the nonlocal parameter goes to zero.

4.3.2 Mindlin's Strain-Gradient Beam Equation

The transverse displacement of the Mindlin's strain-gradient beam equation is calculates by following the same procedure and utilizing the four boundary conditions on the clamped end, as shown in Eqs. (4.62–4.65). That is,

$$w = \frac{1}{EI}Q\delta_{sg}^4 - \frac{1}{2EI}Q\delta_{sg}^4\left(e^{\frac{1}{\delta_{sg}}x} + e^{-\frac{1}{\delta_{sg}}x}\right) + \left(\frac{1}{4EI}QL^2 + \frac{1}{2EI}Q\delta_{sg}^2\right)x^2$$

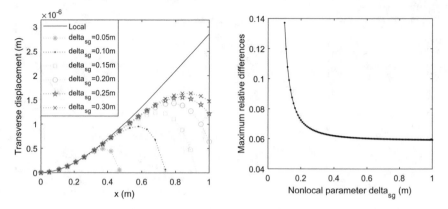

Fig. 4.19 Transverse displacement of the cantilever beam governed by the Mindlin's strain-gradient beam equation. **a** Transverse displacement at different nonlocal parameters, **b** Maximum nonlocal-local differences

$$-\frac{1}{6EI}QLx^3 + \frac{1}{24EI}Qx^4 \tag{4.71}$$

The transverse displacement at different nonlocal parameter δ_{sg} is shown in Fig. 4.19. It can be observed that the cantilever beam governed by Mindlin's strain-gradient beam equation is much stiffer, this is consistent to the deformation of simply-supported and clamped–clamped beam. Also, the nonlocal transverse displacement is close to the local transverse displacement at the clamped end but diverges towards the free end. Moreover, reducing the nonlocal parameter does not make the nonlocal transverse displacement converge to the local deformation. On the contrary, increasing the nonlocal parameter make the nonlocal deformation closer to the local deformation but not converge. This can be explained by the exponential terms in Eq. (4.71).

4.3.3   Higher-Order Beam Equation

The transverse displacement of the higher-order beam equation is

$$w = \frac{Q}{EI}\left(\delta_{sg}{}^4 - \delta_{sg}{}^2\delta_{eg}{}^2\right) - \frac{Q}{2EI}\left(\delta_{sg}{}^4 - \delta_{sg}{}^2\delta_{eg}{}^2\right)\left(e^{\frac{1}{\delta_{sg}}x} + e^{-\frac{1}{\delta_{sg}}x}\right)$$
$$+ \left[\frac{1}{4EI}QL^2 + \frac{Q}{2EI}\left(\delta_{sg}{}^2 - \delta_{eg}{}^2\right)\right]x^2 - \frac{1}{6EI}QLx^3 + \frac{1}{24EI}Qx^4 \tag{4.72}$$

The transverse displacements at different nonlocal parameter δ_{sg} and δ_{eg} are shown in Fig. 4.20. It can be observed that increasing δ_{sg} or decreasing δ_{eg} would make the cantilever beam much stiffer, while decreasing δ_{sg} or increasing δ_{eg} would make the cantilever beam much softer. This is consistent to the observations in the simply-supported and the clamped–clamped beams. Moreover, the higher-order beam reduces to the local Euler Bernoulli beam when $\delta_{sg} = \delta_{eg}$, this can be confirmed by the Relative nonlocal-local displacement differences shown in Fig. 4.21.

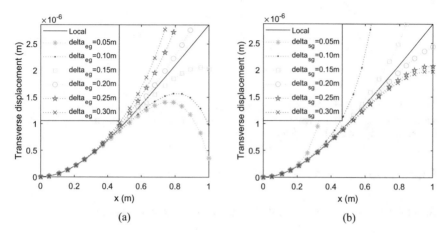

Fig. 4.20 Transverse displacements of the cantilever beam governed by higher-order beam. **a** the transverse displacements at $\delta_{sg} = 0.18$; **b** the transverse displacements at $\delta_{eg} = 0.18$

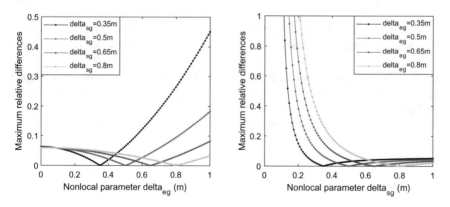

Fig. 4.21 Relative nonlocal-local displacement differences of the simply-supported beam to the nonlocal parameter

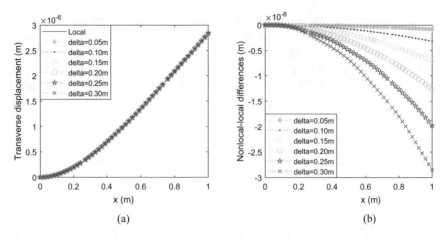

Fig. 4.22 Transverse displacement of the cantilever beam governed by the truncated 4th order peridynamic beam. **a** Transverse displacement at different nonlocal parameters, **b** nonlocal-local differences along the beam at different nonlocal parameters

4.3.4 Truncated Peridynamic Beam Equation

Then the transverse of the truncated peridynamic cantilever beam is

$$w(x) = \left(\frac{QL^2}{4EI} - \frac{Q\delta^2}{72EI} \right)x^2 - \frac{QL}{6EI}x^3 + \frac{Q}{24EI}x^4 + \frac{Q\delta^4}{1296EI}\left[1 - cos\left(\frac{6x}{\delta} \right) \right].$$
$$(4.73)$$

Then the deformations of the cantilever beam subjected to different values of nonlocal parameter are drawn in Fig. 4.22. It can be observed that the cantilever beam governed by truncated peridynamic beam equation is much stiffer. This is different from the observations in the simply-supported and the clamped–clamped beams. Moreover, both the analytical expression in Eq. (4.73) and the Fig. 4.22 show the fact that the transverse displacement of truncated peridynamic beam reduces to the local Euler Bernoulli beam as the nonlocal parameter goes to zero.

References

1. Eringen AC (1972) Linear theory of nonlocal elasticity and dispersion of plane waves. Int J Eng Sci 10:425–435
2. Eringen AC (1983) On differential equations of nonlocal elasticity and solutions of screw dislocation and surface waves. J Appl Phys 54(9):4703–4710
3. Peddieson J, Buchanan GR, McNitt RP (2003) Application of nonlocal continuum models to nanotechnology. Int J Eng Sci 41:305–312

4. Challamel N, Reddy J, Wang C (2016) Eringen's stress gradient model for bending of nonlocal beams. J Eng Mech 142(12):04016095
5. Mindlin RD (1963) Micro-structure in linear elasticity. Arch Ration Mech Anal 16(1):51–78
6. Mindlin RD (1965) Second gradient of strain and surface-tension in linear elasticity. Int J Solids Struct 1:417–438
7. Lim CW, Zhang G, Reddy JN (2015) A higher-order nonlocal elasticity and strain gradient theory and its applications in wave propagation. J Mech Phys Solids 78:298–313
8. Diyaroglu C, Oterkus E, Oterkus S, Madenci E (2015) Peridynamics for bending of beams and plates with transverse shear deformation. Int J Solids Struct 69–70:152–168
9. Diyaroglu C, Oterkus E, Oterkus S (2019) An Euler-Bernoulli Beam formulation in an ordinary state-based peridynamic framework. Math Mech Solids 24(2):361–376
10. Wang Q, Shindo Y (2006) Nonlocal continuum models for carbon nanotubes subjected to static loading. J Mech Mater Struct 1(4):663–680
11. Wang Q, Liew K (2007) Application of nonlocal continuum mechanics to static analysis of micro- and nano-structures. Phys Lett a 363:236–242

Chapter 5
Numerical Solutions to Peridynamic Beam

This chapter first discusses the peridynamic boundary condition treatments. Then, numerical solutions to the integral-form peridynamic beam equation is solved. Benchmark examples, including the simply-supported beam, the clamped-clamped beam and the cantilever beam, are solved with pseudo-layer boundary condition treatments.

5.1 Peridynamic Boundary Condition Discussion

The peridynamics has problems on boundary due to the lack of material filling in the horizon of the boundary nodes as shown in Fig. 3.5. As a result, the boundary region is much softer compared to the interior region [1]. And the boundary induced problems are often called 'surface effect'. Traditional treatment to remedy the surface effect is by adding a pseudo-layer outside the physical boundary with certain thickness, as shown in Fig. 5.1. This thickness usually equals to one horizon size for bond-based peridynamics and to two horizon size for state-based peridynamics [2]. The displacement on the pseudo-layer is defined properly considering the physical property of the boundary.

One way to assign the displacements on the pseudo-layer is by extrapolating functions over the physical domain. The deformation of pseudo-nodes is then determined by the extrapolation function, while the coefficients of the extrapolation function is calibrated from the deformation of physical nodes [2]. For example, the odd function is defined over the boundary to describe pseudo-layer displacement for the Dirichlet-like conditions, and the even function is applied over the boundary for the Neumann-like condition [3]. Accordingly, the peridynamics kernel functions can be extended on boundary with even and odd functions [4]. Similarly, pseudo-layer particles can be enriched with deformation obtained by the symmetric rule [5] or by the linear extrapolation function [6]. The matching boundary condition

J. Chen, *Nonlocal Euler–Bernoulli Beam Theories*,
SpringerBriefs in Continuum Mechanics,
https://doi.org/10.1007/978-3-030-69788-4_5

Fig. 5.1 Pseudo-layer
boundary condition
enrichment [2]

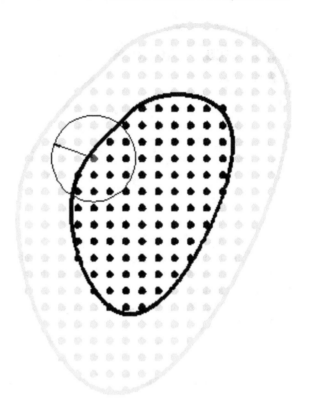

approach obtains the displacement and velocity of the boundary nodes by extrapolating both displacement and velocity of physical nodes [7–9]. After utilizing the
iterative process, the displacement and velocity of pseudo-nodes can be obtained
from the matching condition [10]. The nonlocal boundary condition can also be
represented by localizing influence function near the boundary with a recoverable
convex kernel approximation [11, 12].

In addition to the traditional pseudo-layer boundary condition treatment, other
recently developed boundary condition treatments require no pseudo-layer enrichment, including the volume correction method [13], the force density correction
method [14], the energy correction method [14], the force normalization method
[15], the positive-aware linear solid constitutive model [16] and the fictitious nodes
method [17]. Moreover, the peridynamic equilibrium equation can be reformulated
as an infinite order differential equation by Taylor's series expansion, then, the equilibrium equation on boundary is formulated by truncating the differential equation
up to the second order [18, 19]. Thus, the formulated governing equation enables
the direct imposition of nonlocal boundary conditions. A comparative study on the
efficacy of these surface effect reduction methods is proposed by Le and Bobaru
[20]. Recently, a boundary condition treatment called variable horizon approach is
proposed [2]. The size of the horizon is assigned depending on its distance to the

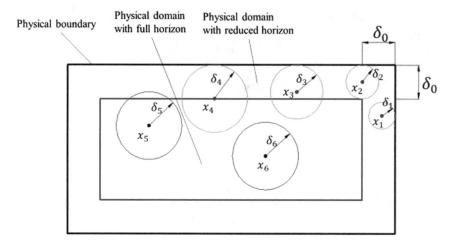

Fig. 5.2 Variable horizon configuration

boundary, as shown in Fig. 5.2. Region close to the boundary within distance equals to one time horizon is assigned with reduced horizon value for bond-based peridynamics. The interior region remains full horizon. The efficacy of this boundary condition treatment is discussed by one- and two-dimensional benchmark examples.

5.2 Numerical Solution to Benchmark Examples

As previously derived in Chap. 3, the integral-form peridynamic beam equation can be expanded into infinite order differential equation. And the analytical derivation in Chap. 4 shows that the truncated 4th order differential equation exhibits nonlocal effects. Since the integral-form peridynamic beam equation is a Fredholm integral equation and its analytical solution can not be derived directly at current stage, the integral-form peridynamic beam deformation are numerically solved. Three benchmark examples are calculated with the particular boundary condition configuration, including the simply-supported beam, the clamped-clamped beam and the cantilever beam.

The discretized beam is shown in Fig. 5.3, where the pseudo-regions are added outside the physical boundary to represent the integral-form peridynamic boundary conditions. The circle nodes represent the physical domain, which are numbered

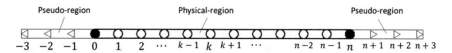

Fig. 5.3 Discretization of the beam

from 0 to n. The triangle nodes are pseudo-nodes, which are added outside of the physical boundary to represent the nonlocal boundary condition.

Then the integral-form peridynamic beam Eq. (3.26) is discretized into

$$M_i = \frac{EI}{\delta} \sum_{\substack{j = i - k \\ j \neq i}}^{i+k} \frac{w_j - w_i}{(x_j - x_i)^2} \Delta x_j, \tag{5.1}$$

where j is the number of the node located within the neighbor of node i, k is the number of nodes located within one side of the neighbor of node i. After expanding Eq. (3.26) at $i = 1, 2, 3, \ldots, n - 1$ and formulating the displacement of nodes on pseudo-region, the transverse displacement can be calculated.

5.2.1   Peridynamic Simply-Supported Beam

The simply-supported beam subjected to a uniformly distributed load is shown in Fig. 4.1. The nonlocal boundary conditions are formulated in Eqs. (4.3)–(4.6). The integral form boundary conditions are defined on the pseudo-nodes outside the physical boundary. The displacement relation between the pseudo-nodes and the physical nodes depends on the property of the constraint on the boundary and requires calibration. In this example, the peridynamic boundary conditions are defined at

$$x = 0:$$

$$w_0 = 0 \tag{5.2}$$

$$w_{-k} = -w_k \tag{5.3}$$

and $x = L$:

$$w_n = 0 \tag{5.4}$$

$$w_{n-k} = -w_{n+k}, \tag{5.5}$$

where the lower indices $k = 1, 2, 3 \ldots$ depends on the value of δ/dx.

The transverse displacement of the simply-supported beam governed by the integral-form peridynamic beam equation is shown in Fig. 5.4. The norm of the local-peridynamic difference as a function of nonlocal parameter δ is shown in Fig. 5.5.

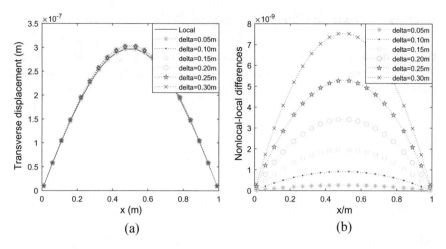

Fig. 5.4 Transverse deformation of the simply-supported beam governed by the integral-form peridynamic beam equation. **a** Transverse displacement; **b** local-nonlocal differences

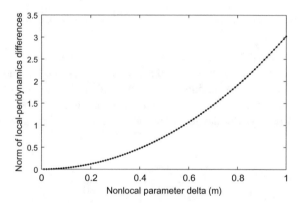

Fig. 5.5 Norm of the local-peridynamic difference of simply-supported beam

It can be observed that the integral form peridynamic beam deformation reduces to the local beam deformation as the nonlocal parameter δ goes to zero. The nonlocal effect becomes severe at higher nonlocal parameter and reduces the beam's stiffness since the beam with higher nonlocal parameter deforms larger. However, the transverse displacement has no fluctuation at certain horizon value and the norm of the local-peridynamic difference increases asymptotically with the increase of the nonlocal parameter. This is due to the fact that the truncated 4th order beam equation in Chap. 4 is a coarse approximation of the integral form peridynamic beam equation. The higher-order terms in peridynamic beam equation cancel the fluctuation of the transverse displacement.

5.2.2 *Peridynamic Clamped-Clamped Beam*

The clamped-clamped beam subjected to a uniformly distributed load is shown in Fig. 4.10. The nonlocal boundary conditions are formulated in Eqs. (4.39)–(4.42). The pseudo-nodes displacements are defined at

$$x = 0 :$$

$$w_0 = 0 \tag{5.6}$$

$$w_{-k} = w_k \tag{5.7}$$

and $x = L$:

$$w_n = 0 \tag{5.8}$$

$$w_{n-k} = w_{n+k}, \tag{5.9}$$

where the lower indices $k = 1, 2, 3 \ldots$ depends on the value of δ/dx.

The transverse deformation of the clamped-clamped beam governed by the integral-form peridynamic beam equation is shown in Fig. 5.6. The norm of the local-peridynamic difference as a function of nonlocal parameter δ is shown in Fig. 5.7. Similar to the deformation of the simply-supported beam, the nonlocal

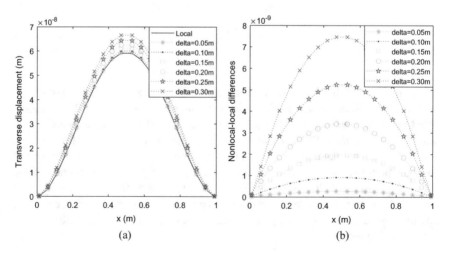

Fig. 5.6 Transverse deformation of the clamped-clamped beam governed by the integral-form peridynamic beam equation. **a** Transverse displacement; **b** local-nonlocal differences

Fig. 5.7 Norm of the local-peridynamic difference of clamped-clamped beam

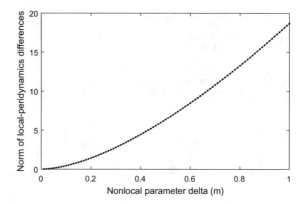

effect reduces the beam's stiffness at higher values of nonlocal parameter δ. Also, the higher-order differential terms in the integral-form peridynamic beam equation suppress the fluctuation of the transverse displacement.

5.2.3 Peridynamic Cantilever Beam

The cantilever beam subjected to a uniformly distributed load is shown in Fig. 4.17. The nonlocal boundary conditions are formulated in Eqs. (4.62)–(4.67). The pseudo-nodes displacements are defined at

$$x = 0 :$$

$$w_0 = 0 \tag{5.10}$$

$$w_{-k} = w_k \tag{5.11}$$

and $x = L$:

$$w_{n-k} + w_{n+k} = 2w_n, \tag{5.12}$$

where the lower indices $k = 1, 2, 3 \ldots$ depends on the value of δ/dx.

The transverse deformation of the cantilever beam governed by the integral-form peridynamic beam equation is shown in Fig. 5.8. The norm of the local-peridynamic difference as a function of nonlocal parameter δ is shown in Fig. 5.9. It can be observed that the peridynamic cantilever beam is stiffer than the local beam. The numerical transverse displacement calculated from the integral-form beam equation is close to the truncated order peridynamic beam equation shown in Fig. 4.22.

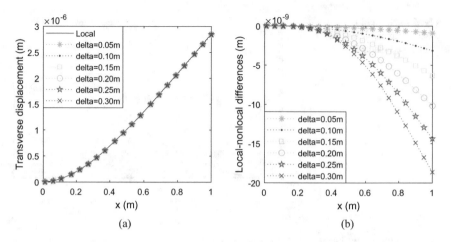

(a)   (b)

Fig. 5.8 Transverse deformation of the cantilever beam governed by the integral-form peridynamic beam equation. **a** Transverse displacement; **b** local-nonlocal differences

Fig. 5.9 Norm of the local-peridynamic difference of cantilever beam

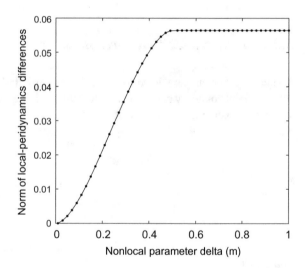

References

1. Silling S (2000) Reformulation of elasticity theory for discontinuities and long-range forces. J Mech Phys Solids 48:175–209
2. Chen J, Jiao Y, Jiang W, Zhang Y (2020) Peridynamics boundary condition treatments via pseudo-layer enrichment method and variable horizon approach. Math Mech Solids. https://doi.org/10.1177/1081286520961144
3. Zhou K, Du Q (2010) Mathematical and numerical analysis of linear peridynamic models with nonlocal boundary conditions. SIAM J Numer Anal 48(5):1759–1780
4. Aksoylu B, Celiker F (2017) Nonlocal problems with local Dirichlet and Neumann boundary conditions. J Mech Mater Struct 12(4):425–437

5. Yaghoobi A, Chorzepa MG (2018) Formulation of symmetry boundary modeling in non-ordinary state-based peridynamics and coupling with finite element analysis. Math Mech Solids 23(8):1156–1176

6. Shojaei A, Mossaiby F, Zaccariotto M, Galvanetto U (2018) An adaptive multi-grid peridynamic method for dynamic fracture analysis. Int J Mech Sci 144:600–617

7. Nicely C, Tang S, Qian D (2018) Nonlocal matching boundary conditions for non-ordinary peridynamics with correspondence material model. Comput Methods Appl Mech Eng 338:463–490

8. Wildman RA, Gazonas GA (2011) A perfectly matched layer for peridynamics in one dimension. Army Research Laboratory Technical Report, Adelphi. MD (2011) ARL-TR-5626

9. Wildman RA, Gazonas GA (2012) A perfectly matched layer for peridynamics in two dimensions. J Mech Mater Struct 7(8–9):765–781

10. Wang L, Chen Y, Xu J, Wang J (2017) Transmitting boundary conditions for 1D peridynamics. Int J Numer Methods Eng 110:379–400

11. Wu CT (2014) Kinematic constraints in the state-based peridynamics with mixed local/nonlocal gradient approximations. Comput Mech 54:1255–1267

12. Wu CT, Ren B (2015) Localized particle boundary condition enforcements for the state-based peridynamics. Coupled Syst Mech 4(1):1–18

13. Bobaru F, Foster JT, Geubelle PH, Silling SA (2017) Handbook of peridynamic modeling. CRC Press, Taylor & Francis Group, Boca Raton

14. Madenci E, Oterkus E (2014) Peridynamic theory and its applications. Springer, Berlin

15. Macek RW, Silling SA (2007) Peridynamics via finite element analysis. Finite Elem Anal Des 43:1169–1178

16. Mitchell JA, Silling SA, Littlewood DJ (2015) A position-aware linear solid constitutive model for peridynamics. J Mech Mater Struct 10:539–557

17. Oterkus S (2015) Peridynamics for the solution of Multiphysics problems. PhD thesis, The University of Arizona

18. Chen J, Tian Y, Cui X (2018) Free and force vibration analysis of peridynamics bar with finite boundary conditions. Int J Appl Mech 10:1850003

19. Madenci E, Dorduncu M, Barut A, Phan N (2018) Weak form of peridynamics for nonlocal essential and natural boundary conditions. Comput Methods Appl Mech Eng 337:598–631

20. Le QV, Bobaru F (2018) Surface corrections for peridynamic models in elasticity and fracture. Comput Mech 61:499–518

Chapter 6
Conclusion

In this book, the deformation of Euler–Bernoulli beam has been studied comparatively by utilizing different nonlocal beam theories, including the Eringen's stress-gradient beam equation, the Mindlin's strain-gradient beam equation, the higher-order beam equation and the peridynamic beam equation. All these nonlocal theories introduce one or two nonlocal parameters for the Euler–Bernoulli beam equations. Benchmark examples are solved analytically using these nonlocal beam equations and are compared to their local counterparts, including the simply-supported beam, the clamped–clamped beam and the cantilever beam. Results show that these nonlocal beam equations reduce to the local Euler–Bernoulli equation when the nonlocal parameter goes to zero. For the simply-supported beam and the cantilever beam, the Eringen's stress-gradient beam equation and the peridynamic beam equation yield a much softer beam deformation, while the beam governed by the Mindlin's strain-gradient beam equation is much stiffer. However, the cantilever beam deforms otherwise. Moreover, the Euler–Bernoulli beam governed by the higher-order beam equation can be stiffer or softer depending on the values of the two nonlocal parameters. Furthermore, a fluctuation on the displacement of the truncated-order peridynamic beam equation is observed on the simply-supported and the clamped–clamped beam and is explained from the singularity of the solution expression. Therefore, the deformation of the beam depends on its governing equation as well as its boundary conditions. Finally, the integral-form peridynamic beam equation is solved numerically after constructing certain boundary conditions. This comparative study on nonlocal beam theories intends to impress the readers on the complicated behaviors of Euler-Bernoulli beams and the importance of nonlocal boundary condition treatments.

J. Chen, *Nonlocal Euler–Bernoulli Beam Theories*,
SpringerBriefs in Continuum Mechanics,
https://doi.org/10.1007/978-3-030-69788-4_6

Printed in the United States
By Bookmasters